DID THOM ▮▮▮▮▮ REALLY INVE ?

INVENTIONS THAT CHANGED
OUR HOMES AND OUR LIVES

CATHERINE O'REILLY

SKYHORSE PUBLISHING

Skyhorse Publishing books may be purchased in bulk at special discounts for sales promotion, corporate gifts, fund-raising, or educational purposes. Special editions can also be created to specifications. For details, contact the Special Sales Department, Skyhorse Publishing, 555 Eighth Avenue, Suite 903, New York, NY 10018 or info@skyhorsepublishing.com.

www.skyhorsepublishing.com

10 9 8 7 6 5 4 3 2 1

Library of Congress Cataloging-in-Publication Data

O'Reilly, Catherine.
Did Thomas Crapper really invent the toilet? : inventions that changed our homes and our lives / Catherine O'Reilly.
p. cm.
Includes bibliographical references and index.
ISBN 978-1-60239-347-9 (alk. paper)
1. Household appliances—History. 2. Inventions—History. I. Title.
TX298.O73 2008
643'.6—dc22
2008025969

Printed in China

*"Men say they know many things
But lo! They have taken wings,
The arts and the sciences,
And a thousand appliances...."*

—Henry David Thoreau [1817–1862]

*"I just let electricity do my work nowadays. I have an electric
dishwasher, ...electric clothes washer, ...electric iron, ...electric
vacuum, ...toaster stove, ...coffee percolator."*

—Housewife in a Western Electric Company ad, 1915

Contents

CONTENTS

CONTENTS

CONTENTS

CONTENTS

THOMAS CRAPPER & COMPANY, LTD.

SANITARY ENGINEERS

By Warrant of Appointment to their Late Majesties Edward VII and George V.

ESTABLISHED 1861.   INCORPORATED 1904.   COMPANY No. 82482.

Introduction

I'm sure Thomas Crapper would be intrigued to find his name on the cover of this delightful work. I am the managing director of his firm, Thomas Crapper & Co. Ltd., (established 1861), and I was honored to be asked to write this introduction.

Mr. Crapper richly deserves his fame, but I have to admit that oftentimes he is credited with too many innovations. Sadly, he did not invent the W.C., the cistern, the syphon, or the water trap, but he was much more than a simple plumber. Crapper had his own factory which manufactured his exclusive wares, and in addition, he had some of his exclusive designs produced by other firms, who were specialists in their fields.

Crapper was a great businessman, sanitary reformer, and self-publicist. He relentlessly promoted sanitary fittings to a skeptical world; up until then, bathroom suppliers and manufacturers were discreetly situated in side streets. Mr. Crapper introduced the first bathroom showroom: imagine the fuss when Crapper & Co. opened in London's King's Road, opposite Royal Avenue, with W.C. bowls in the windows! Ladies fainted in the street.

Thomas Crapper's inventiveness was well known; he registered a number of patents, one of which was for the "Disconnecting Trap," which became an essential underground drains fitting. This alone was a great leap forward in the campaign against disease. On the other hand, there was one for a *spring-loaded* W.C. seat, which as

THE STABLE YARD,   ALSCOT PARK,   STRATFORD-ON-AVON,   WARWICKSHIRE.   (CV37 8BL)

ELECTRIC MESSAGES: wc@thomas-crapper.com   TELEPHONE: ALDERMINSTER (01789) 450 522.
MANAGING DIRECTOR – S.P.J. KIRBY.   FACSIMILE: "   "   "   "   523.

INTRODUCTION

the user stood up, swung sharply to an upright position, pulling rods that automatically flushed the cistern. Unfortunately, over time the rubber buffers on the underside of the seat would perish and become sticky. This caused the seat to remain down as the incumbent rose. Moments later, under stress from the powerful springs, the seat would free itself and sweep violently upwards—striking the unfortunate Victorian on the bare bottom! The device—not a commercial success—became popularly known as "The Bottom Slapper."

Thankfully, such setbacks were few, and by the 1880s, Crapper & Co.'s reputation was such that they were invited to supply the Prince of Wales (later Edward VII) at Sandringham. Subsequently, Windsor Castle, Buckingham Palace, and Westminster Abbey all benefited from Crapper goods and services. Today, the Crapper manhole covers in the abbey are popular with tourists for wax rubbings! Crapper & Co. remained by Royal Appointment to Edward when he became king and was also warranted by George V, as Prince of Wales and once again as king. The company was held in great respect and it prospered: the first choice of royalty, nobility and gentry.

No single individual "invented" the loo, and the slang word "crap" was not derived from Thomas Crapper's name. (It was an old English word that fell out of use in Britain, but continued as a slang word in America and Canada). Yet the two are forever linked in our minds. During World War I, American servicemen stationed in London were so amused that the ancient and vulgar word for faeces was printed on so many water closets, that they began to call the W.C. a "Crapper." Though crude, the soubriquet made sense and it stuck. Therefore, "crap" is an old word but "Crapper" comes from our man, Thomas Crapper. In etymological circles, this process is called a "back-formation," which sounds rather like a sewer problem!

Aren't you glad you asked?

Simon Kirby,
Thomas Crapper & Co. Ltd.,
Stratford-on-Avon,
Warwickshire, England.

Low-Level Cistern.

SUPPLIED PRIMED -
CAN BE PAINTED
ANY COLOUR.

No. 814 (L)

THE FOLLOWING ARE BRIEFLY SOME OF THE NUMEROUS
FEATURES OF EXCEPTIONAL MERIT
TO WHICH CONSIDERATION IS RESPECTFULLY DIRECTED.

AUTHENTICITY - The very best of the Edwardian low-level cisterns sported this type of "beer-engine" handle, so we are proud to re-introduce this pattern.

QUALITY - Constructed and finished in the same way as our high level cistern but fitted with a china-handled lever. Available with Brass, Nickel-Plate or Chromium-Plate fittings.

We recommend that it is used in conjunction with our brackets and patrases, large-bore low-level flushpipe, and oval wooden seat (see illustration, left).

A PROPER <u>SYPHON</u> IS SUPPLIED - NEVER A "FLUSHING VALVE."

BRITISH MADE.

Bathroom

ARCHAEOLOGISTS HAVE FOUND THAT BASIC PLUMBING systems date back more than five thousand years: Copper pipes were found beneath an Indian palace from around 3300 BCE. The ruins of the ancient Romans' *thermae* (public baths) indicate that bathing was a public activity in a society that did not consider nudity exceptional, particularly during public sports and recreational events. Although the ancient Romans were the first to use baths for physical cleanliness, some anthropologists feel that human bathing originated from religious rituals. Plumbing and the installment of bathtubs in individual homes did not come until as late as the nineteenth century.

Although there is still controversy about the modern bathtub's true origins, John Michael Kohler seems to hold the inventor's reins. Kohler's bathtub began as a four-legged, enamel-lined horse trough, but evolved into a claw-footed bathtub. There are also tubs that use pedestals rather than claw feet. On the island of Crete, an ancient five-foot long pedestal tub was found. Fast-forward to the high-style tubs of today, some of which are recessed into the floor or a platform.

Descriptions of past cultures mention times when bodily dirt and body odors were masked with perfumes and cosmetics, rather than being removed by washing. That's not the case today, as most Americans subscribe to the notion that "cleanliness is next to godliness." Lucky for us, bathtubs abound.

KITTY LITTER FIRST ENTERED THE LEXICON OF CAT LOVERS in the 1950s when entrepreneur Edward Lowe, whose family owned an industrial absorbents company in Minnesota, provided a neighbor with absorbent clay called Fuller's earth to replace the ashes she was using in her litter box. He went on to make a fortune under the brand name "Kitty Litter."

Fuller's earth is any fine-grained, naturally occurring earthy substance that has a substantial ability to absorb impurities. Its name originated with the textile industry, in which textile workers (or fullers) cleaned raw wool by kneading it with a mixture of water and fine earth that absorbed oil and other contaminants from the fibers.

Before 1950, most cat boxes were filled with sand, dirt, or ashes, so the advent of a highly absorbent substance that also didn't leave a mess throughout the house was a great boon for cat owners. However, odor remained a problem, particularly if the litter wasn't replaced in a timely fashion. The bacteria found in the cat's feces converts the uric acid in cat urine into a noxious ammonia odor that is all too often associated with cleaning out the litter.

The advent of clumping materials, more effective in moisture absorption, helped solve this. In the 1980s, Thomas Nelson, an American biochemist, developed a commercially viable clumping litter from bentonite clay.

IN 1872, A FRENCH HAIRDRESSER BY THE NAME OF Marcel Grateau came up with the idea of using heated tongs to wave hair to create the Marcel Wave. This clever invention would later be known as a curling iron.

In 1906, a permanent wave machine was demonstrated for the first time in London. Its purpose was to make a woman's hairstyle last longer. Specifically, Karl Ludwig Nessler, a hairdresser from Germany, was able to use his device to set the Marcel Wave more permanently, using a combination of borax paste and electrically-heated curlers. It was known as the Nestlé permanent wave.

Nessler's method worked, but was expensive and uncomfortable—the brass curlers weighed more than a pound each and the technique took many hours. Yet, in spite of its problems, the permanent wave machine became the rage in America after Irene Castle (1893–1969), the famous ballroom dancer, used it for her high fashion curled hairstyle during her dancing career in the 1920s. Curled bobs became a symbol of affluence in popular movies during this time period.

As technology improved, hair curling irons became sleeker and easier to use. You can now buy them in the local drugstore for less than $20.

WHEN DID MOST PEOPLE BEGIN WEARING DEODORANT? The answer, my folk-song singing friend, is blowing in the wind. In the animal kingdom, odors are used to attract or repel. The advertising world does not want word of this leaking out, though! Body odor is whole-heartedly discouraged on our civilized streets.

Masking one unwanted scent with another more-appreciated scent has been the general method behind the first deodorants for more than five thousand years. Every major civilization has left a record of its efforts to produce deodorants. The early Egyptians, Greeks, and Romans relied on scented baths and perfumed oils.

Ironically, though human perspiration is itself mostly odorless, it wasn't until scientists discovered the true origin of body odor—when moisture is fermented by bacteria that thrive in hot, humid environments (such as underarms), it starts to smell—that we were able to virtually eradicate body odor. Deodorants are now usually alcohol-based and work at neutralizing the growth of bacteria in addition to the masking scent. Then the antiperspirant was born: aluminum-based salts, such as aluminum chloride, actively prevent our underarms from sweating in the first place.

The first deodorant was introduced in 1888. It was called Mum. Mum was not exactly what we're used to finding in a deodorant today. It sold by the jar, as a cream that you smeared on with your fingers.

It wasn't until the late 1940s that a researcher at Bristol-Myers unveiled Ban Roll-On. The inspiration came from another new invention with a completely different purpose: the ballpoint pen.

Aerosol deodorants were invented in the 1960s, obviously before we realized we were harming the earth's ozone layer. No matter. Sweaters are still staying cool and dry with today's variety of stick, roll-on, and non-CFC aerosol deodorants.

Has your dentist recently recommended that you floss? He would not have been the first. It's been nearly two hundered years since Levi Spear Parmly, a dentist from New Orleans, invented dental floss from silk thread. It wasn't until around World War II that Dr. Charles C. Bass developed a nylon floss which was found to be better than silk because of its greater elasticity and resistance to abrasion. The first company to patent dental floss was Johnson and Johnson in 1898.

Generally, floss is dispensed in a small plastic container with anywhere from ten to fifty feet of nylon floss. Nylon floss is available waxed and unwaxed, and in a variety of flavors. Because this type of floss is composed of many strands of nylon, it may sometimes tear or shred, especially between teeth with tight contact points. Single filament (PTFE) floss slides easily between teeth, even between tight spaces, and is virtually shred-resistant.

Alternatively, one-hand flossers have been developed to allow for greater ease in flossing. These are small plastic devices with a short strip of nylon floss stretched between two small posts attached to a small handle.

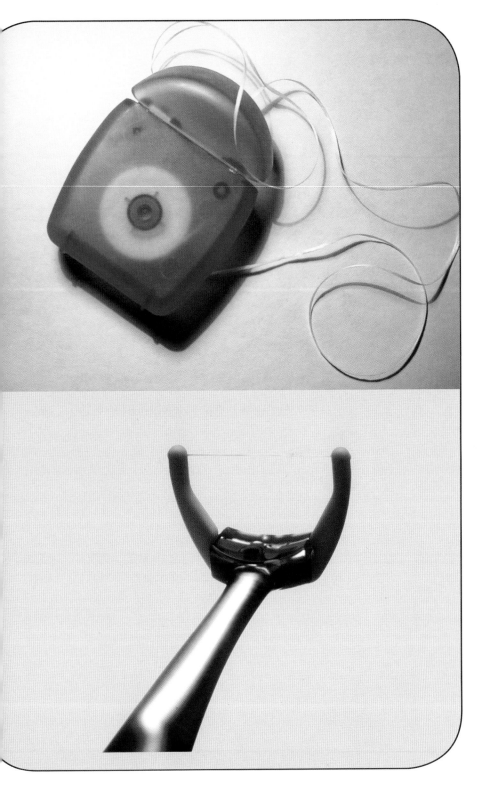

IN 1996, AN MIT SURVEY NOTED THAT THERE WERE AS many Americans who said that they "could not live without a blow dryer," as those who said the same about their personal computer.

The original concept for the hair dryer started as a secondary selling point for early vacuum cleaners. As advertised, not only would they clean your dirty floors, but if the flow of the warm air were reversed, the vacuum could also dry hair more efficiently. However, the large and heavy household device needed some tweaking to make it appropriate for hair care. Alexandre Godefoy invented the first electric hair dyer in 1890, but it was nearly as large and cumbersome as the vacuum-dryer. The first hand-held hair dryer was introduced in the early 1900s. Hollywood movies set in the 1920s sometimes used the hair blower, as well as the white telephone, as a symbol of a luxury only the leisure class could afford.

During the Roaring Twenties and up to World War II, many hair-styles did not need to be styled before the hair dried. The hooded dryers were popular in hair salons, but not in typical homes. A compromise for hands-free styling at home could be achieved by placing a hand dryer in a holding clip on a stand.

ALTHOUGH STYLES HAVE CHANGED, THE QUEST FOR BEAUTY has remained a consistent pursuit of cultures throughout the centuries. Methods for "enhancing" skin have been occasionally painful, soothing, creative, and ridiculous. In order to achieve soft, unwrinkled, and radiating skin, men and women used, and still use, lotion.

When he discovered Tutankhamun's resting place in 1922, Howard Carter found more than gold in the pharaoh's tomb. A little bottle of lotion lay beside the Egyptian king, presumably in order to keep his skin eternally smooth and healthy in the after life. But this skin remedy dates even beyond King Tut. Also in Egypt, preserved tablets dating as far back as 3000 BCE provide in depth beauty concoctions and tricks to avoid wrinkles, dry skin, and blemishes. For example, to combat aging, Egyptian women were encouraged to use a creamy paste of milk, wax, crocodile dung, olive oil, and juniper leaves.

Lotion continued to grace the skin of men and women throughout the centuries. In England during the Elizabethan era, women desiring a glazed look applied a paste made of egg whites to their faces. About 200 years later, during the Regency Period, the British considered pale skin to separate the wealthy from the poor, as the former had no need to labor in the sun. Women wore bonnets and powders, but also began using lotions and makeup with white lead and mercury to whiten their complexions. Unfortunately, these products were extremely toxic, resulting in illness and sometimes death.

As a result of the fatal effects of homemade lotions, women started to rely more heavily on marketed items. In 1880, the lumberjack Andrew Jergens formed a company with a soap maker. The two created a product that hit the market and flew off the shelves. In 1899, George Bunting formulated his own lotion. Originally marketed as "Dr. Bunting's Sunburn Remedy," his lotion was used by men and women to soothe burned skin and even treat conditions such as eczema. First referred to by satisfied customers as "No-eczema," his brand Noxzema still flourishes today.

Today, the lotion market is packed with "perfecting" products. With items claiming to de-wrinkle ageing faces, firm the butt, and provide healthy and glowing skin, women and men still rush to counters, looking for skin remedies with as much fervor as in the 1800s.

BELIEVE IT OR NOT, EVEN PRIMITIVE MAN SHAVED HIS stubble by using sharpened flint and shell (the original razors). Ancient cave drawings unearth images of both bearded and clean-shaven cave men.

Like many cosmetic developments and trends such as ghost-white makeup or bound feet, hair and hair removal became a symbol of status and proper maintenance for many centuries and in cultures such as ancient Greece, Egypt, and Rome.

The gradual discovery of metal allowed razors to evolve from stone to copper, iron, and bronze. These altered shavers had either wooden or thick metal handles.

By the sixteenth century, razors were individually fashioned by the local blacksmiths, and resembled hatchets. These were the first straight razors, which would dominate the market until the early 1900s.

In 1740, Benjamin Huntsmann of Sheffield England changed the razorblade forever. He developed a method to produce purified steel, stronger and safer than any of the previously used metals. Attached to wood and bone handles—or ivory for the classy shaver—these razors were used to remove men's whiskers until 1900. In the age of the straight razor, shaving was considered an art, perfectly accomplished by only the best barbers. However, thanks to the mind of King Gillette, the safety razor now resides in individual bathrooms, relatively knick free.

The royalty of the razor world, King Camp Gillette of Wisconsin, took the development of the shaver to the cutting edge. The new throw-away bottle caps inspired Gillette's imagination. He conceived of a safe and cheap disposable razor with a protective edge and thin steel blade, as opposed to the sharp edges of the widely used straight razor. Six years later, Gillette's razor was patented in 1901 and he was a millionaire by 1910.

Another man obsessed with a good clean shave, Jacob Schick patented the first electrical razor in 1923. He devised a handheld, motor-powered device for shaving, but the product inconveniently

required both hands—one to hold the motor and the other to hold the attached razor. His invention needed work. Finally, after a motor small enough to fit inside the razor was invented, the first electric dry shaver was introduced in 1931 by Schick Incorporated in Stamford, Connecticut.

THE WORD "SHOWER" CAN REFER TO MANY DIFFERENT things, but in household parlance it is a shortened reference to the "shower bath." The dictionary definition for that phrase is "a bath in which water is sprayed on the bather from an overhead nozzle."

But the nozzle was not always necessary. In the past, servants filled up buckets with water, slowly emptying them over their wealthy masters. For centuries, this remained the most common shower. However, a few shower-like devices did exist. While excavating an ancient Egyptian tomb in the city of Tel-el-Amarna, researchers discovered a basin designed specifically for standing and showering. Servants still acted as nozzles, but this tub is considered one of the first showers. In ancient Greece, people would bathe under the spouts of public fountains, allowing the plumbing to do all the work. But the modern shower didn't develop until the nineteenth century.

The American Virginia Stool shower from the 1830s was one of the earliest forms of showers. Picture this: an all-wood contraption with a revolving seat, a hand-operated water pump, and a foot-pedal controlled scrub brush. That was the American Virginia Stool Shower. Not very luxurious, but it certainly got the job done.

Although Russian and Turkish bathhouses boasted hot tubs, showers, and steam rooms, the early standard American shower-er seemed to be content with a single showerhead.

The development of several shower accessories has helped showering become a safe and pleasant experience. They include hand showers, body sprays, shelves, radios, and even television devices, allowing you to boogey down while you wash up.

THE CONCEPT OF TEETH CLEANING HAS BEEN AROUND since the first chewing sticks of 3500 BCE (see Toothbrush), but the pioneers of teeth whitening were the barbers of the Middle Ages. The barbers would file down their customer's teeth and then apply a coat of strong acid. Ultimately, this method caused the enamel to decay, ruining teeth, but boy did they look white right after the acid was applied!

As toothbrushes and toothpaste became increasingly common in the eighteenth and nineteenth centuries, the acidic wash used through the 1700s decayed in popularity. Ammonia, which originally entered the whitening department early on in the form of human urine, became a main ingredient in toothpastes, along with fluoride. Although urine is left out of modern toothpastes, the ammonia and fluoride remain.

Today, the teeth whitening business flourishes and there are a variety of different methods and products on the market. Not surprisingly, teeth whitening is close to a two-billion-dollar-a-year industry.

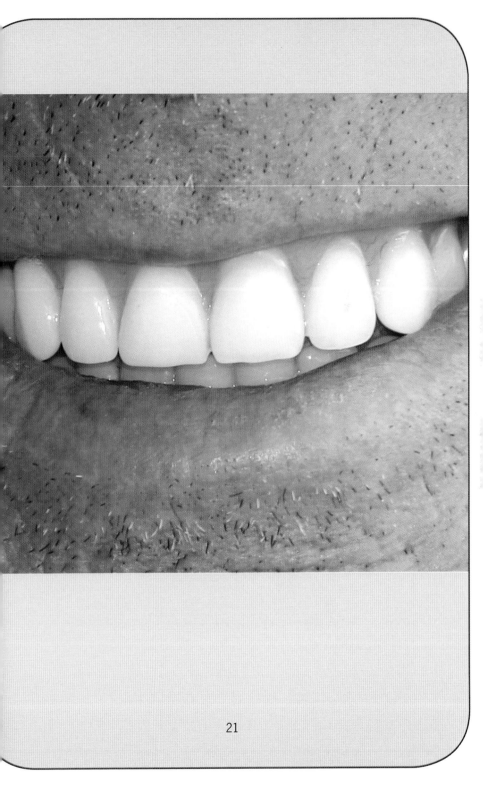

THE DAYS OF CHAMBER POTS AND OUTHOUSES GAVE way to a more modern toilet in the eighteenth century in England. However, back in 1596, Sir John Harington, the poet and godson of Queen Elizabeth I, invented a toilet water closet that apparently had most of the basic features of today's restroom, even a flush toilet, albeit non-electric. Harington installed one in Richmond Palace, his royal godmother's home, but his invention was largely ignored by the rest of society.

The first mechanical indoor toilets seem to have made their appearance in the 1730s, but the first patent was granted much later (1775) to Alexander Cummings, a London watchmaker. Cummings improved upon Harington's flush toilet by adding a water trap (or "u-bend") that kept sewer gases from leaking into the room.

In 1870s London, water closets offered only on-off valves on their tanks, much as does a typical sink. Because huge amounts of water were being wasted, the local administration called for the introduction of a "Valveless Water Waste Preventer." Thomas Crapper heavily promoted a special flushing mechanism for the tanks (called a "siphon") which, unlike a valve, simply cannot leak. He manufactured tens of thousands of tanks, usually with the company name emblazoned on the front. The siphon soon became the only legal flushing device in Britain until 2001, when, under pressure from Europe, flushing valves were legalized. A pity really, as valves can leak and waste water if the seal fails. Thomas Crapper & Co. is still trading, now making exact replicas of its Victorian toilets, but on principle (and in honor of the founder) they never supply flushing valves—only siphons!

There you have it. Thomas Crapper did not invent the toilet. Mr. Crapper, however, did hold nine plumbing-related patents that improved the toilet (improvements to drains, water closets, manhole covers, and pipe joints). The Crapper name most likely became synonymous with the toilet because "Crapper" was marked proudly on the tanks and toilet bowls and because he made and installed them for princes and kings!

As FAR BACK AS 3500 BCE, THE BABYLONIANS' VERSION of the toothbrush was known as a "chewing stick." Think: an aromatic stick the size of a pencil, with one end to chew until it becomes soft and bristled and the other end pointed to pick food out from between your teeth.

In the fifteenth century, Europeans began toting around the first toothbrush invented by the Chinese. Made of stiff Siberian wild boar hair and a bamboo or bone handle, the first toothbrush was not comfortable to put in one's mouth. Until Wallace H. Carothers invented nylon in the twentieth century, tooth brushers would replace boar hair with other animal hair.

The first mention of toothbrushes in England was in a letter dated 1649. In the letter, someone traveling to Paris was asked to bring back what he described as being very similar to today's most simple basic non-electric brush. About fifty years later another letter mentioned brushes being sold by J. Barret, a local merchant. In the 1700s, sets of brushes were sold, each brush being a different size.

H. N. Wadsworth patented the first toothbrush in 1857, and in 1885, American companies began to mass-produce these toothbrushes. While toothbrushes became widely available, brushing habits were just being developed. It wasn't until soldiers brought their toothbrushing duty back home, that Americans began to pick up on regular dental practices.

SAID THE ACTRESS (WHO DRESSES WELL) ABOUT THE chest of drawers in a bedroom (a dresser) and a cupboard for kitchen dishes (a dresser) owned by the person (a dresser) who cares for her clothes: "My dresser's dresser has a dresser in her bedroom and another in her kitchen."

But what about the strange lady who keeps dishes in her dresser? The unusual use of the word "dresser" as a cupboard comes from the origins of the Middle English word *dressour*, a table for preparing food.

The most common use of the word refers to a set of drawers for storing clothes. It is the bedroom furniture known as a chest of drawers or, in North American English, as a bureau. It has rows of one, two, or a combination of columns of drawers. A combination of wardrobe and chest of drawers is called a chifforobe (a term that combines the words chiffonier and wardrobe). A chiffonier is a narrow, high chest of drawers, often with a mirror. If a tall chest of drawers is mounted on legs, it is called a highboy. Its cousin, lowboy, is a table-like chest of drawers. Variations of the word date back to the homes of medieval nobility in Europe.

The variety of available chests of drawers should make it possible to find exactly what fits the envisioned space, and the things to be stored, the desired décor, as well as the budget.

BED WARMERS FOR BED AND LAP BLANKETS, OFTEN UTILIZED by people traveling in buggies, were used even in stoic colonial America. The device was simply a flat, pan-shaped iron covered with thick cloth that could retain heat for a while after being in an oven.

Heating pads were introduced in Europe for personal use in 1912. Other applications of electricity to enhance personal comfort were much more elegant in the many advanced details of the doomed ocean liner Titanic, such as its electric elevators. Meanwhile, for people of ordinary means, a hot-water bottle costing fifty-nine cents in 1914 and heavy wool blankets costing one dollar and sixty-eight cents would keep one warm at home and safe from icebergs.

In 1927, the first "warming pad" was marketed in Great Britain, by Thermega Ltd., London.

The first automatic electric blanket was invented in 1936, in the wake of the Great Depression, when few ordinary households in the United States could afford it. In fact, an electric blanket with automatic controls was still a luxury item for most Americans in 1946. At that time, one type was selling for thirty-nine dollars and fifty cents.

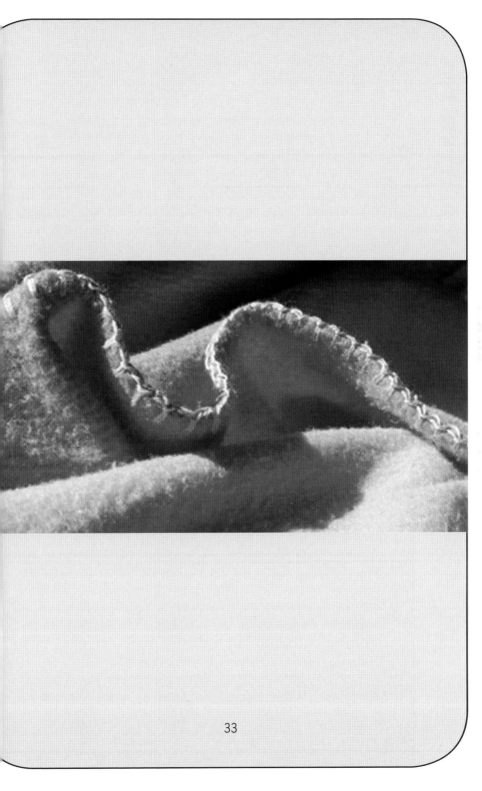

ANYONE WHO HAS BEEN SUPPORTED GENTLY WHILE peacefully floating on something secure and comfortable might have discovered the delights of water bedding. In 3500 BCE, when the common mattress consisted of animal skins hand-stuffed with grass and reeds, the Persians looked for a more luxurious option. The first water beds, goat skin filled with water, carried them off into a river of dreams. The water bed didn't officially surface until centuries later.

British physician Dr. William Hooper developed the first model of the water bed out of rubber in the late 1800s presumably to help prevent the kind of bedsores that can result when invalids rest in one position too long. Hooper's waterbed didn't fare too well—it constantly leaked.

Water beds didn't gain popularity until about the 1970s, right when durable, watertight vinyl came around.

In 1968, nearly a century after Dr. Hooper's efforts, Charles Hall and other design students at San Francisco State University made a successful water bed—by mistake. Originally, Hall just wanted to invent an innovative chair. The first prototype for the chair was a vinyl bag filled with 300 pounds of cornstarch. The results were not comfortable. Failure after failure prompted Hall and his students to eventually throw their chair endeavors out the window and rest on a water *bed* (literally). Once the water bed was perfected, the watery vinyl wonder became hugely popular.

A patent was granted in 1986 for a "water bed" that is actually an air mattress that floated on water. "Free flow" mattresses today have only one water chamber and need time to stabilize after being disturbed. Mattresses that combine interconnected air and water chambers are called "waveless" water beds.

Cleaning Tools

A BRIEF REVIEW OF NOTES FROM CHEMISTRY 101 MIGHT help clean things up while trying to understand laundry bleaches.

Bleach helps detergents get rid of soils and stains by converting dirt to more soluble particles that can then be removed by a detergent. The dirt is carried away in the wash water. Before throwing out the wash water, the proverbial baby should be removed first, of course.

The active ingredient of bleach is sodium hypochlorite. Sodium hypochlorite was first discovered by a French chemist named Claude Louis Berthollet in 1785. Berthollet's discovery was followed by a Scottish chemist named Charles Tennant, who and developed a bleaching powder that grew popular throughout Europe.

The brand Clorox is nearly synonymous with laundry bleach, and for good reason. The Electro-Alkaline Co. (now known as Clorox Co.) produced and sold large quantities of bleach—but only to groups like commercial laundries and water companies in the early 1900s.

History might have swept under the rug forever the name of the witch who first constructed a broom and then flew off to sweep the floor of her castle. But she will probably be pictured most often with a corn broom, rather than a push broom.

Historic creations of early eighteenth-century American homes include handmade brooms. Wood and brush as found in nature were used. Corn husks or stiff weeds were tied onto a stick. Even a century later, the simple tool is basically the same as those listed in the 1897 Sears Roebuck & Co. catalog, which offered sixteen kinds of brooms. The company pledged, "All our broom handles are thoroughly seasoned. The brooms are warranted not to come off the handle." Then house brooms cost two dollars . . . a dozen!

Corn brooms are the most basic and familiar domestic sweeping tool, but heavier-duty versions for dealing with heavy debris or snow might have a steel band holding together a corn exterior and a tougher yucca fiber center. Some brooms have washable plastic fibers.

In 1905, in Hartford, Connecticut, a Nova Scotia man named Alfred Carl Fuller constructed wire and bristle brushes by night. By day, Fuller was an impressive salesman, selling his brushes door-to-door at fifty cents each. By 1910, he had a staff of twenty-five to manufacture and sell the famous brush. The "Fuller Brush Man" retired in 1943, with annual sales topping ten million dollars.

WHAT COULD BE WORSE THAN LOOKING THROUGH streaky windows? This was the solution that Windex®—the first brand of glass cleaner—brought to the problem of cleaning glass. No more streaks! Windex seems to have delivered on its promise as it remains the most popular of glass cleaners and has entered everyday language as the generic reference for any brand of glass cleaner.

Invented in 1933 by Harry R. Drackett, Windex was essentially 100 percent solvent, and, because it was flammable, had to be sold in metal cans. The product was reformulated after World War II, when modern surfactants were introduced. Surfactants are wetting agents that lower the surface tension of a liquid, allowing easier spreading and are soluble in both organic solvents and water (See: Fabric Softener).

Before the widely recognized blue liquid cleaned glass surfaces, there was . . . water. Water mixed with alcohol or vinegar was arguably the first glass cleaner solution. Simple yet effective.

Today, glass cleaners typically contain 90 to 95 percent water, 1/10 to 1 percent by weight ethylene glycol n-hexyl ether, and 1 to 5 percent isopropanol by weight. And among the chemicals we are actually familiar with: detergents (to decrease surface tension of the solvent) and ammonia (one of several chemicals that is used to quicken drying time, therefore reducing streaks, residue, and spots).

THE FAMILIAR HOUSEHOLD OBJECT CALLED A MOP HAS been so called for centuries, although in recent years the word applies to a variety of cleaning implements that our grandmothers would not recognize. They are all made of absorbent material fastened to a handle and used for dusting, and washing and drying floors or other usually flat surfaces.

These modern and often specialized tools are likely to phase out the old-fashioned mop someday, leaving future generations to wonder why a loosely tangled bunch or mass is called a mop of hair.

Originally, the basic wet mop was simply a bunch of yarn attached to a long broom handle. It achieved greater fame when paired with the wringing mechanism attached to a pail, which was invented and patented in 1893 by Thomas W. Steward. This made it easy for the user to wring out and then renew the water in the mop.

There are wet mops, sponge mops (some with built-in wringers), and dust mops. Since 1995, static cleaners with disposable paper or cloth covers or nylon mops have been used to gather loose dust and small objects.

Hammacher Schlemmer offers Scooba, a robotic floor washer that "works without human intervention."

THE WORLD HAS BEEN BUBBLING OVER WITH SOAPY SUDS since 600 BCE, when the Phoenicians created a cleansing substance by boiling animal fat, water, and ash and then allowing it to evaporate, leaving a waxy detergent. When excavating the ancient city of Pompeii, covered in ash and lava in 79 CE, an entire soap factory was discovered beneath the hardened surface. This hygienic item flowed in and out of popularity with the trends of hygiene and, oddly enough, religion.

In the Middle Ages, the Church encouraged cleansing of the soul, but not necessarily of the flesh. Fire-and-brimstone lectures on the evil of exposing skin stalked Christians while they bathed, and soap fell out of favor along with the exposure of flesh. Soap prevailed once more as bacteria and dirt were linked with disease and infection. People rushed to the stores to supply their families with the saving suds.

Some of the new arrivals from England to Jamestown, Virginia, in 1608 were soap makers. They had practiced their craft at home for years, apparently not marketing their practical products beyond their small circle of contacts. Meanwhile, in France, advances in soap making were made, first by Nicholas Leblanc in 1791, then by Michel Eugène Chevreul in 1811. Leblanc prepared for modern soap making by using his patented process for making high-quality sodium carbonate. Chevreul's later work laid the foundation for all soap chemistry.

As is true of many of the incredible innovations in this book, one special form of soap was created entirely by accident. Following the lead of William Colgate, William Proctor and James Gamble started their own company in 1837, mass-producing soap and candles to compete with the imported and luxurious soaps popular at the time. Proctor's son, Harley, suggested they make a lightly scented and delicately white soap. This product soon floated to success as a result of a happy accident. Breaking for lunch, one of the workers forgot to turn off the giant kettle heating and stirring the soon-to-be-soap, allowing too much air into the mixture. That day, the floating soap later known as "Ivory" was born.

IN THE MID-NINETEENTH CENTURY, SEVERAL PATENTS FOR such sweepers for home use were given in the United States. Their typical use was on carpets, where they were often the least effective.

In 1858, there was the carpet sweeper. It cleaned without suction—it was just a rotating brush that used wheels to activate it—but eleven years later, the first suction vacuum cleaner arrived. Manufactured and sold by the American Carpet Cleaning Company, this suction vacuum cleaner was powered by a suction fan driven by a hand crank.

In 1881, homemakers rolled out the red carpet for an invention that used bellows to create a draft that helped pick up loose dust that was then swept up by revolving brushes.

Bissell, a familiar name in the industry, refers to Melville Reuben Bissell, who, in 1876, patented a better carpet sweeper than any previously. A million Bissell sweepers were produced annually by 1906.

In 1901, an engineer named Hubert Cecil Booth watched workers use compressed air to move dust around while trying to clean a carriage at a London train station. He later declared that he thought the dust should be collected in a receptacle rather than be left free to be blown around yet again. He built a prototype and formed the Vacuum Cleaner Company Limited in 1902.

The first upright vacuum cleaner with the familiar dust bag attached to the handle was built by a department store janitor in the city now called Canton, Ohio. James Murray Spangler made his machine from "a broom-handle and an old pillow case begged from his wife for a dust-bag."

Spangler's simple contraption won the attention of an Ohio harness maker named W. H. Hoover. His main business was being swept away by the growing popularity of the horseless carriage. He made a clean sweep of his own business and needed something new and in demand. He bought Spangler's rights and, in 1908, the electric carpet sweeper was born. Soon and for decades later his name would be nearly synonymous with the vacuum cleaner. By 1912, electric floor cleaners were very popular and several models were offered in mail-order catalogues.

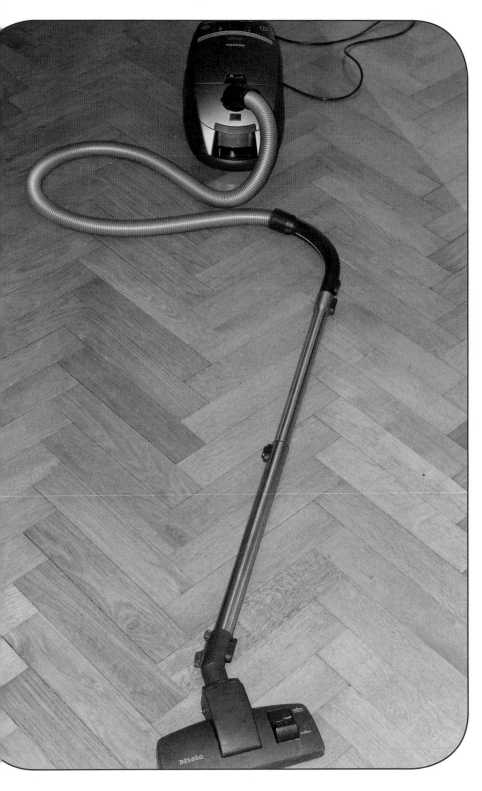

House

DR. JOHN GORRIE CONSTRUCTED THE FIRST AIR-conditioning contraption in 1844, at the American Hospital for Tropical Fevers in Apalachicola, Florida. Today's basic cooling systems, like refrigeration, are loosely modeled on Gorrie's invention—that is, "cooling caused by the rapid expansion of gases." The original air conditioner was actually used to treat yellow fever, as Dr. Gorrie saw healing properties in cold air. In 1851, Gorrie received a U.S. patent, and in 1855, he died before having the chance to further market his invention.

In 1902, William Haviland Carrier installed an air-conditioning system at a Brooklyn printing plant. Four years later, he obtained a patent for his apparatus, and went public in 1911.

A household version of Carrier's cooling system appeared in 1928; however, few could afford such an expensive appliance during the economic depression of the 1930s. After World War II, the interest in domestic air-conditioning was renewed in the United States, but generally not throughout Europe, where it is still not as popular as in the States.

Overall, domestic and exported unit shipments of home comfort appliances, room air conditioners, and dehumidifiers doubled between 1983 and 1995.

ACCORDING TO THE CARPET AND RUG INSTITUTE, THE first woven carpet mill in the United States began in Philadelphia in 1791. In the early 1900s, mills also opened in New England and New Jersey. The first famous name in carpets dates back to 1839, when Erastus Bigelow invented a power loom for weaving carpets. With his machine (which is now on display in the Smithsonian), production doubled the first year and tripled by 1850.

Bigelow introduced the first broadloom carpet in 1877. The next year, four Shuttleworth brothers established their plant in Amsterdam, New York. Nearly twenty-five years later, they introduced the Karnak Wilton carpet. Tufted plastic carpeting first appeared in 1953, not to be confused with the artificial turf used later on some football fields or miniature golf courses.

Physicians, environmental professionals, consumer groups, carpet installers, and carpet makers (including Monstanto and DuPont) are among those concerned about the Environmental Protection Agency's policies and actions concerning toxic carpet and indoor air quality. Concerned parents have brought issues to the attention of nursery schools, athletic programs, and others who make decisions about the carpeting used where children work or play.

IN 1882, DR. SCHUYLER SKAATS WHEELER WAS THE main wheeler dealer (aka chief engineer) of the Crocker and Curtis Electric Motor Company, New York. That was the year he developed the first electric fan. The manufacturing of two-bladed desk fans began the next year. Wheeler's two-bladed electric desk fan with "no protective cage" was inspired by the scientific work of the great Thomas Edison and Nicola Tesla.

Meanwhile, Philip H. Diehl, an employee of the Singer Sewing Machine company, constructed the first ceiling fan by mounting a fan blade to a sewing machine motor and then mounting the contraption to the ceiling. Thus, the first ceiling fan was born. Diehl patented his invention in 1887.

The ceiling fan would eventually take on innovative changes with the time's scientific revolution. Diehl later added lights to the new ceiling fan and reduced the size of the fan's motor.

During the twenty years following Wheeler's first fan, the alternatives for cooling off with electric fans either blew in one direction or revolved. In 1908, the Eck Dynamo and Electric Company developed the first gear-driven oscillating electric fan, giving the user more control over the area to be cooled and cutting down the waiting time for the fan to return to cool the user directly.

THE DEHUMIDIFIER COMES TO THE RESCUE IN HOMES WITH excess moisture and condensation. Humid air can cause mold and mildew to grow, which can lead to health problems. The dehumidifier sucks the humidity through a screen and deposits it as water in a dish that can be emptied regularly.

In the 1950s, a company called Munters (dubbed the "humidity expert") introduced the idea of an air-conditioning system "based on the use of evaporative cooling and dehumidification." This would become the heart of today's modern dehumidifier.

The most common types of dehumidifiers are the mechanical, or refrigerative, dehumidifiers. A fan draws in moist air and passes it over a refrigerated coil. However, the colder the climate, the less effective the appliances might be. Another type is a desiccant dehumidifier. The process is complicated, but, to simplify, a fan moves the air through the desiccant holder, another fan moves low-humidity air through the holder, and a heater prepares the air that dries the desiccant.

Fortunately, users need not be engineers to simply turn the device on and benefit from the scientists' understanding of it all.

As soon as fire was discovered, man started using it for warmth. But it took a very long time for the fireplace to become what it is today. Surrounding stones and a makeshift vent or hole that allowed smoke to escape skyward were the primary characteristics of the primitive fireplace. Poorly ventilated, they provided a big serving of warmth, soot, ash, and indoor air pollution.

In the 1700s, when fireplaces were predominantly made out of plaster or stone, Abraham Darby began using new methods of smelting (the method of making new metals). He soon discovered iron and its resiliency to heat, as well as its ability to radiate and withstand extreme temperatures without chipping or cracking. It was also around this time that fireplaces began to gain aesthetic value in the home. The shifting role of the fireplace along with the discovery of new metals gave way to blacksmiths and stonemasons to create a new breed of more attractive, efficient fireplaces.

Benjamin Franklin further improved the fireplace (and woodstove) in the 1740s, with his invention of the freestanding cast-iron stove. The heavy iron Franklin stove kept heat radiating throughout a room even after a fire had gone out. There was just one problem: The Franklin stove vented smoke from the bottom of the unit, preventing the smoke from rising and not allowing the unit to work properly.

By the late 1780s, a Philadelphia resident named David R. Rittenhouse redesigned the fireplace by adding an L-shaped pipe behind the unit that would act as a chimney.

While Franklin was busy improving the fireplace, another Benjamin—Benjamin Thompson, the Count of Rumford—designed a fireplace that is now the foundation for modern fireplaces. By making them smaller, but taller, more shallow, and with wider angles, heat could radiate even better. Today, we know his design as the Rumford fireplace.

WHAT DO WICKS SOAKED IN ANIMAL FAT, OIL LAMPS OF Egypt, wax candles from Rome, and the gaslights of England, all have in common? They provided light. But all required fire. With the introduction of electricity in the nineteenth century, artificial light would continue its evolution into the lightbulb.

Throughout the 1800s, electricity provided the curious minds of inventors and scientists with a new frontier to explore. In 1802, Sir Humphrey Davy created a few minutes of light by sending an electrical current through a strip of platinum, provoking a chemical reaction to give off electricity. He invented the Arc lamp in 1809 by connecting two charcoal sticks and an early battery. The electrical current passed through the carbon created by the vaporized charcoal, creating an arch-shaped glow across a four-inch gap. The light was small, but extremely bright. However, the lamps were difficult and expensive to mass produce.

Thomas Edison of America and Joseph W. Swan of England entered the effort to create long-lasting and cheap electric light in the second half of the nineteenth century. They would invent the incandescent lightbulb. Swan began experimenting with carbon paper filaments within a glass bulb, but he would not succeed in creating a working, patentable product until 1878. After testing various metals, Edison carbonized a piece of sewing thread, which produced a lasting electrical current, and in 1879, he succeeded in creating a bulb that burned for thirteen and a half hours. He continued to experiment with the size of the carbon filament and the electrical current, improving upon his own and Swan's lightbulb to eventually create a bulb lasting for 1,200 hours.

Since then, the lightbulb has gone through many improvements. In 1906, General Electric patented a method of making tungsten filaments to use in incandescent lightbulbs. William David Coolidge further enhanced this method in 1910 by inventing a method to make tungsten filaments that lasted longer than other types. Coolidge's method was also less costly. About twenty years later, Marvin Pimpkin introduced "soft" light by patenting a

process for frosting glass interiors, and in 1936, the more familiar coiled coil filament was finally introduced.

The recent switch to energy-saving fluorescent lighting and the distinctive curly-tube bulb can trace its origins back to the fluorescent lumiline lamp developed by General Electric at Nela Park, Ohio. It was a 24-inch long tube that gave off a green light. It was first shown at the annual convention of the Illuminating Engineering Society in Cincinnati, Ohio, in 1935. The first public use of the new lighting was at a celebration of the centenary of the U.S. Patent Office in 1936.

The fluorescent lamps of both GE and Westinghouse launched their competitive products the very same day: April 1, 1938. No fooling.

LONG BEFORE "BE COOL" REFERRED TO A CERTAIN contemporary style, it indicated a burning desire for one to make himself comfortable in hot weather in stressful social situations. Ladies especially achieved this coolness by fanning themselves with a fan, a little object now practically gone from daily life. The fan was in fact an integral part of a lady's costume. The handheld fan became an often elaborate object of art.

Where there's hot weather, there's always a desire to be cool. The first mechanical fans were found in the hot Middle East. These fans, developed in the nineteenth century, were called "punkah fans." The punkah was comprised of a rope and pulley system. Servants would pull on the rope suspended from the ceiling, moving the giant punkah leaf up-and-down to cool their masters. The start of the Industrial Revolution led to an increase in hot and overcrowded factories and mines, and the need to be cool increased with them.

In 1900, General Electric addressed this necessity, albeit not the need to send subtle social signals, by concentrating on harnessing electrical power rather than human hand movements. It was one of the first projects undertaken after its research laboratory was founded in 1900. Only two years later, James J. Wood, a consulting engineer, received a patent for an electric fan produced at the Fort Wayne, Indiana, Electrical Works.

THE HEALING QUALITIES OF LIGHT HAVE BEEN PRESCRIBED for centuries. In ancient Greece, Hippocrates encouraged many of his patients to expose themselves to light to cure their ills. Now, artificial light therapy in the form of UVA and UVB rays from lamps is used to treat diseases such as eczema and seasonal affective disorder.

In 1904, the German company Heraeus invented the first high-pressure UV lamp. On their Web site they bask in the glory as "the founder of tanning with synthetic light sources," claiming the light output of the first UV lamp has "virtually the same quality as the sun." This lamp was primarily used to help sick patients, but in 1978 a German professor named Friedrich Wolff recognized an alternative use for the UV lamp.

While running a study on the effects of UV light on athletes, Wolff found one he wasn't expecting. Their skin turned a fashionable golden brown, and soon after, Wolff developed the first tanning beds. Since then, tanning has become an incredibly successful industry, especially in Europe and the United States. Until 1988, there were no restrictions on the usage of tanning beds or tanning lamps, but when serious health issues were brought up in relation to the beds, numerous health organizations cracked down on tanning salons.

No ONE KNOWS WHO FIRST INTENTIONALLY PASTED PAPER to wall. Whoever it was undoubtedly caused a spouse to ask, "You think you're going to do *what* to these walls?"

Perhaps we can look to Egyptian papyrus, dating back six thousand years, to find wallpaper's origin. Its decorations might appear to be a primitive comic strip, but actual history and commentary was recorded with such symbols and was drawn on the walls of great public buildings at the time.

In 200 BCE, the Chinese glued rice paper to their walls. The first wallpaper? Could be. At any rate, the art of papermaking spread throughout Asia and Europe for centuries.

During the eighteenth century, when wall decorations began to include paper, the first wallpapers were sought after because they resembled the more expensive wall treatments of that time: embossed leather, wood paneling, and tapestry. When Christophe-Philippe Oberkampf invented the first wallpaper printing machine in 1785 and Louis Robert created the "endless roll," wallpaper started a slew of printing, design, and pasting advances.

Plastic resins appeared in the mid-1940s. As wall covering, they excelled paper in nearly all ways: durability, strength, flexibility, and maintenance.

"HONEY, BE A DEAR AND RUN ME A BATH!"

But without the Englishman Maughan's invention of the first instant water heater in the 1870s, the only way to "run" a bath was to boil the water in a large kettle and pour it into the tub. His invention then influenced Edwin Rudd, a Norwegian mechanical engineer, who invented the automatic storage water heater in 1889 in Pittsburgh.

The basic modern water heater consists of a large steel tank, holding forty or sixty gallons, with an inner glass liner to keep rust out of the water. The tank is wrapped in insulation. A dip tube lets cold water enter the tank and another pipe lets hot water out. A thermostat controls the temperature inside the tank by controlling when the electrical elements in the tank or the gas-fired heater at the bottom turn on and off. A drain valve allows the tank to be emptied for repairs or moving, and a pressure relief valve keeps the tank from exploding.

Present-day hot water offerings include tankless and solar heaters. Solar water heaters consist of a solar collector, which converts solar radiation into usable heat; a heat exchanger/pump, which transfers the heat from the solar collector into the water; and a storage tank to store the solar-heated water.

Kitchen

IN 1916, A COMMERCIAL BLENDER WAS PATENTED IN America and was used especially for mixing drinks, a very popular activity even during the infamous Prohibition era (1920–1933), during which the manufacturing and distribution of alcoholic products went underground.

As soda fountains gained popularity, the need for better blending appliances was seen and gradually satisfied, first by Hamilton Beach, and later to be coupled with Proctor-Silex. In the 1940s, Oster Manufacturing, later to become part of Sunbeam, and Waring were the successful trendsetters in the industry.

In 1924, Hamilton Beach introduced the first soda fountain mixer for making milkshakes. In 1932, an electric juicer named Juicit was introduced by the Chicago Electric Company. Four years later, the Waring blender was produced, soon to become a very recognizable national brand name. A rival machine, the Juice-O-Mat, was made by Rival Company in the 1950s.

Blenders were limited by the number of accessories they could boast, which led to the more squat and versatile food processor. In addition to blending, these machines could shred carrots, mix pie crust, and chop nuts. The first successful American version of the pioneering unit was introduced in 1973 and was called the Food Processor by Cuisinart.

IN THE LATE EIGHTEENTH CENTURY, A FRENCH CHEF named Nicholas-François Appert developed tinplate canisters designed especially to preserve food. In 1810, a patent was taken out in England by Peter Durrand for the use of "vessels of tin or other metals" for preserving food. The business partners Donkin and Hall acquired the rights of the patent. They opened the first cannery at Blue Anchor Road, Bermondsey, England, in 1812.

Printing directly onto tin, a lithographic process that began in London in 1853, helped to identify contents, but attaching paper labels to the can remains a more economical solution. Printed tin artifacts, such as antique toys, are especially popular collectables. Louis Pasteur's breakthrough work with bacteria brought attention to the importance of sterilization, which led to reforms in food-processing. The time was right for extensive use of canned food.

The modest can is an unsung essential of modern life. There are self-heating cans, self-cooling cans, and microwaveable cans. Canning is the method of preserving and packaging food, without which civilization never would have ventured beyond the local food supply. It changed the way the world eats and revolutionized the food industry.

IF CHEESE IS GREAT, ISN'T GRATED CHEESE GREATER? Apparently so, because nearly every well-equipped kitchen has at least one of several kinds of cheese grater.

An aboriginal people called the Ngadjonji once used a flat stone with deep grooves cut into its surface to grate yellow walnuts and zamia nuts. The grated nuts would make up a type of flour that was baked on stones and then eaten. We now know the grater as a utensil used to reshape certain foods into smaller pieces, such as crumbs or strips. Citrus peel (especially of lemons, limes, or oranges) can be turned into zest.

One of the lesser-known uses comes from tropical regions, where cheese graters are used on the meat of coconut to make the familiar small slivers of coconut that decorate and add flavor to cakes. In Jamaica, objects shaped like cheese graters or even the cheese grater itself, are used as percussion instruments. They are like cowbells or other objects that can produce a distinctive sound when struck.

THE FRENCH *BIGGIN*, A SIMPLE CONTAINER MADE UP OF two metal containers separated by a plate with holes that acts as a filter, was created in 1800.

The coffee percolator was invented in 1806 in Germany by the American Benjamin Thompson (known as Count Rumford). He knew of and improved on the Turkish method of heating ground beans and water together. An alternative method was to let gravity pull filtered water from the upper part of a coffeemaker, through the coffee, and into the lower part of the unit.

The first coffee machines were the large but simple drip machines that appeared in public eating places after 1830. Later, and into the early years of the twentieth century, the machines used electricity rather than steam heat. By 1908, there were electric percolators in England, and then electric models in the United States and Germany that were simply electrically-heated percolators. Also, filtering problems with the early coffeemakers led a woman by the name of Melitta Benz to develop a filter system that would keep the grounds out of the cup.

Instant coffee was a solution to some problems, as Faust brand coffee reminded drinkers in 1919 that there was no waste, no grounds or leaves, no boiling, no pots to clean. Yet, coffeemakers prevailed and improved year after year.

The birth of the automatic drip coffeemaker finally arrived in 1972 under the new brand, Mr. Coffee. The largest coffeemaker manufacturer in the world to date, Mr. Coffee paired the benefits of the automatic drip and disposable filter to permanently simplify the coffeemaking process. In just three years after Mr. Coffee's launch, the company saw sales increase from one thousand coffeemakers to thirty-eight thousand a day.

THE SMALL BUT MIGHTY CORKSCREW HAS PULLED OUT A lot of stops since being invented sometime during the seventeenth or eighteenth century. The first patent was granted to Englishman Samuel Henshall in 1795. Considered to have been developed from the "gun worm," a device used to retrieve wadding and unspent charges from muskets, the corkscrew's early design was a simple T-shape with a helical steel worm protruding at right angles from the center of a handle made of bone or wood.

The Greeks and Romans used corks to seal bottles, but the corks were tapered and easily removed because part of the cork was always left above the rim of the bottle. With the fall of the Roman Empire, corks fell out of use until they were taken up again in England in the sixteenth century.

During the seventeenth and eighteenth centuries the method of manufacturing glass bottles evolved in England from hand blowing to pouring the hot glass into molds. The mass production of uniform glass bottles with straight sides and narrow necks allowed them to be stored and shipped on their sides. This advantage boosted the international trade in wine.

Tighter seals were then required so the bottles would not leak. Cylindrical corks thus evolved that were compressed prior to being forced into the bottle necks. Because of their tighter fit the corks were harder to remove than the earlier, tapered versions and so arose the need for a method of extracting the cork.

Pulling a cork out requires a force of twenty-five to one hundred pounds, depending on whether the cork is moist (bottle lying on its side) or dry (bottle being stored upright). Since its invention, literally thousands of patents have been taken out on new variations of the corkscrew.

STARTING IN THE LATE 1870S, JOSEPHINE COCHRANE OF Shelbyville, Illinois, began ten years of experiments as she envisioned a dishwashing machine. She built several prototypes, including some driven by steam engine. Various steam engines had existed since the seventeenth century, but none applied to dishwashing.

Cochrane's first commercially successful dishwasher design consisted of a wheel attached to a boiler that held separated wire compartments for the dishes. The motorized wheel would turn as soapy hot water squirted up from the boiler and rained down on the dishes.

After patenting her invention in 1886, Cochrane founded her own company to manufacture her invention, and in 1889, her power model dishwasher was commercially produced. Her company would later become KitchenAid. Cochrane's invention was described at the time as being "capable of washing, scalding, rinsing, and drying from five to twenty dozen dishes of all shapes and sizes in two minutes."

Despite the washer's slow initial sales, technology, dishwashing detergent, and a change in women's attitudes catapulted dishwasher sales in the 1950s.

AN ENGLISH CHEMIST, AMBROSE GODFREY, PATENTED the first automatic fire extinguisher in England in 1723. It consisted of a wooden container of fire-extinguishing liquid which contained a chamber of gunpowder. This chamber was connected to a system of fuses which, when ignited, exploded the gunpowder which then spread the solution.

George William Manby, a British Captain, is considered the father of the modern fire extinguisher. It was invented in 1818 and consisted of a copper container of three gallons (13.6 liters) of pearl ash (potassium carbonate) solution within compressed air.

Fires can be extinguished in one of four ways: by cooling, by smothering, by removing the fuel, or by disrupting the chemical chain reaction thus interrupting the flame. These four methods correspond to four types of fires and their matching types of fire extinguishers.

Serious consequences can result from using the wrong extinguisher to fight a fire. If a water-based extinguisher is used on a flammable liquid fire, for example, the fire may flare up, spread, and cause personal injury to the user and others. Electric shock can result if a water-based extinguisher is used to fight a fire in or near electrical equipment.

THE GARBAGE DISPOSAL'S HISTORY BEGAN IN A BASEMENT in Racine, Wisconsin. An architect named John. W. Hammes wanted to make cooking clean-up easier for his wife, so Hammes threw together the initial garbage disposal in his basement workshop. Later, he rented a local machine shop for twenty-five cents per hour to develop prototypes. Eleven years of testing, developing, and prototyping later, Hammes developed the first fully functional garbage disposal in 1927.

Patented in 1933, the garbage disposal emerged on the market five years later with the company he specifically founded to market his invention: InSinkErator. During InSinkErator's first year, the company produced a mere fifty-two disposal units. Dubbed the "electric pig" when it was first introduced, the garbage disposal's slow start was partly due to the fear of causing problems within the sewer systems.

Once the garbage disposal became accepted into American kitchens—following World War II—InSinkErator began operating against full-fledged competitors who caught wind of Hammes' convenient appliance.

KitchenAid, Whirlpool, and General Electric were right on InSinkErator's heels when Emerson Electric acquired the company in 1968. At this time, garbage disposal sales doubled from 1968 to 1973. In fact, in 1973 alone, 2.7 million units were sold by the disposal industry.

MICROWAVE OVENS REVOLUTIONIZED THE FOOD WORLD by offering fast cooking time, push-button simplicity through digital readouts, and turntable convenience. The turntable supposedly helped cook the food more evenly. Many processed food products were then designed for the microwave, while convenience and sleep overtook considerations of the possible reduction in nutritional value as a result of such cooking.

The microwave oven was the product of an accident and a curious mind. In 1946, Dr. Percy Spencer of the Raytheon Corporation was testing a magnetron, a powerful vacuum tube used in radar systems. Spencer was a self-taught engineer from Maine who had never completed grammar school. During the test, Spencer noticed that the chocolate bar he kept in his pocket had melted onto his pants. He understood that the magnetron worked by radiating waves of microwaves, producing heat. Curious, he sent for kernels of popcorn and placed them near the tube. As the kernels blossomed into fluffy and edible popcorn, the idea for a microwave oven was born. Spencer arrived to work the next morning with an egg, prepared to further test the magnetron's capabilities. He and a colleague watched the egg tremble and then erupt, covering his friend in gooey yolk. It had cooked from the inside out, placing extreme pressure on the shell. Spencer's mind worked as rapidly as the egg had cooked, immediately realizing the potential for a new fast-cooking oven.

His first microwaves, although conveniently fast, were inconveniently sized. The giant metal box, complete with water tubes and fans for cooling the magnetron, weighed more than 750 pounds and stood at five-and-a-half feet tall. Because of its size and high cost, the first microwaves stayed within the walls of the food industry, eager to test the multiple uses of the new invention. Aside from cooking food extremely quickly, it was used to defrost meat, roast coffee, and dry out potato chips. Finally in 1947, the "Radarange" hit the market. As the microwave gained popularity, smaller and safer models were made for the household kitchen and well as the restaurant.

In 1490, THE FIRST DOCUMENTED ENCLOSED STOVE made of brick and tile emerged in France. Although the enclosed range didn't gain popularity immediately, bricks became a common material for warming food. Stacked around the kitchen hearth for cooking, they were often used to warm kettles and pots. John Sibthrope of Britain created a coal-burning version of the enclosed range in 1630, but like the French range before it, it did not replace the open flame. John Mott gave coal another go in 1833. Adding ventilation to his product (called the "baseburner") helped the coal burn more quickly. It wasn't until 1802 that George Brody, another Brit, produced a cast-iron range that would sell, becoming a staple in American and British kitchens. Using steam, his invention heated food thoroughly and evenly. The same year that saw the cast-iron range also witnessed the birth of the gas range. That same year in Germany, Frederick Albert Winston designed the first gas range, but due to the dangers of leaking fumes and burning kitchens, his innovation wouldn't be perfected or manufactured until the 1860s.

In 1889, the proprietor of the Hotel Bernina in Samaden, Switzerland, noticed that the hotel's power supply, generated by a local waterfall, was not being used to its fullest potential. He decided to direct the energy to satisfy cooking needs when it wasn't being used for lighting. The result was the first electric oven.

During the second decade of the twentieth century, when electrical power brought new life to even modest homes throughout America, the oven might have been the appliance after the lightbulb that was dramatically new. In 1910, the first electric range was introduced by Hotpoint. A typical oven unit had three settings: high, medium, and low.

In 1922, the AGA cooker was introduced by Gustaf Dalen. This stove, used both for cooking and heating the home, remained on at all times in order to consistently provide heat. But as central heating became popular in homes, ovens and stoves reclaimed their role as primarily culinary. Also, in efforts to reduce the use of electricity, recent models focus on conserving energy while cooking.

Refrigeration began with holes in the earth covered in snow, pouches of food floating in cool rivers, and caves decorated with ice. Eventually evolving into a high-tech machine, the refrigerator has seen its share of change and development.

As early as 1000 BCE, the Chinese cut ice from mountains, storing it for months at a time by insulating it with wood. In ancient Greece and Rome, snow was imported from mountaintops and sold in snow shops. By the eighteenth century, the British used the icehouse, buildings created aboveground or dug into hillsides for the storage of ice. They would be filled during the wintertime with ice chipped from frozen rivers and lakes. The icebox was long the most popular form of refrigeration. These appliances were most often made of wood and looked like an average piece of room-temperature furniture. However, inside on the top shelf, a large block of ice was placed and replaced once depleted, providing contained and cool air for the items within.

Even up to the beginning of the twentieth century, homemakers were still keeping food cool in iceboxes. Large blocks of ice were delivered to the home two or more times a week. Cards would be displayed in a window of the home indicating the number of pounds of ice needed: twenty-five, fifty, and so forth. A strong deliveryman would carry the requested amount to the icebox and put it in place.

The first attempt at artificial refrigeration took place in 1748, pioneered by William Cullen. Dr. John Gorrie received the first U.S. patent for refrigeration in 1851, as he developed a device to help treat and cool his patients suffering from yellow fever. (Dr. Gorrie is attributed for inventing the air conditioner as well.)

The methods and science behind refrigeration continued to develop and evolve. Based on experiments done by Michael Faraday in England, Frenchman Ferdinand Carre introduced the synthetic refrigerant ammonia in 1859, which would be widely used in refrigerators until the 1920s. And in 1876, Carl von Linden

created a refrigerator that liquefied gas, providing the foundation for modern refrigeration.

In 1913, the first electric home refrigerators were in use, and by 1920 they were popular. A motor and a compressor removed heat from the cabinet and kept the interior cool.

WHERE THERE'S SMOKE THERE'S FIRE, AS THE SAYING goes, and so it stands to reason that, with so much danger posed by fire, people would find a way to detect the smoke before they were consumed by the flames that engulfed them. Francis Robbins Upton, an associate of Thomas Edison, did so in 1890 when he invented the first automatic electric fire alarm.

The first smoke detector patented for home use was invented by Randolph Smith and Kenneth House in 1969. They were the first to feature mass-marketed individual battery-powered units that could be easily installed and replaced.

Today there are two types of smoke detectors on the market: photoelectric and ionized. Photoelectric models depend on a beam of light within the unit being interrupted by an accumulation of smoke in a small chamber, thereby triggering an alarm. This method is slightly less effective because it depends on larger amounts of smoke to set off the alarm.

Ionization models use an ionization chamber and a source of ionizing radiation to detect smoke. This is the more common type of smoke detector because it is inexpensive and better at detecting the smaller amounts of smoke produced by flaming fires, as opposed to smoldering smoky fires. A small piece of americium-241 produces a stream of radioactive particles which generates a small electric current. If smoke enters the chamber, it disrupts the current, triggering an alarm.

ELEANOR ROOSEVELT IS MISTAKENLY QUOTED AS SAYING that "women are like tea bags: people don't know how strong they are until they are in hot water." And people don't know how versatile the tea kettle is until they have one at the ready in the kitchen. Tea time isn't the only time for the kettle to show its mettle.

The steam is responsible for the high-pitched whistle of the pot, alerting everyone within miles that it's ready to be tipped over and poured out. The first evidence of a whistling teakettle was found among the ruins of the Mayan people. One spout was used for releasing the water while another gently whistled. But it wasn't until 1921 that New Yorker John Block recognized the whistle of hot water could be music to his ears—and his bank account. While touring a tea kettle factory in Germany, he remembered the high-pitched noise of his father's pressurized potato cooker and shared this memory with the factory. The factory gave it a shot and the kettles flew off the shelves. The next year he introduced his modified invention in Chicago and it seemed like America could no longer settle for a silent kettle.

Kettles were first made of iron, the most practical metal available at the time. Japanese kettles were shaped like bowls and date from 1517. Teapots evolved through design changes as side handles became swinging arched handles and a groove on the lip became a spout. Examples from various Japanese eras are in museums and they are typically beautifully decorated.

The first electric kettle was marketed by Carpenter Electric Manufacturing Company of St. Paul, Minnesota, in 1891. At this time several great improvements for the kitchen were manufactured, including the electric oven and the bottle cap, manufactured for the first time the next year in Baltimore.

LET US TOAST THE ELECTRIC TOASTER. IT HAS PREPARED and, most Americans would say, "improved" bread for millions of people for more than a century. It faithfully brings the scent of freshly-baked bread, without needing all that kneading.

In 1893, the British manufacturer Crompton and Company invented "The Eclipse," the first electric toaster. A slice of bread was placed near each side of the heated coils for the toasting to happen. It had no controlling adjustment—not even an on/off switch.

When pondering the chicken and the egg in the classic problem about "which came first," one may wonder if pre-sliced bread came before the toaster. The answer quickly pops up: the bread slicing machine was needed first. The inventor, Otto Frederick Rohwedder, began working on it in 1912. It took him nearly two decades of slicing before he designed a machine that sliced and wrapped loaves of bread, in effect making it a perfect match for the toaster.

In 1926, the inventor Charles Strite of the Waters Genter Company added a timer to the toaster and a clever pop-up feature, upgrading the appliance to the "must have" household item it is today.

Laundry

POCHON INVENTED THE FIRST CLOTHES DRYER IN FRANCE in 1800. The concept of the first dryer was very similar to what we are used to today (heat and tumble). Pochon's model—a circular metal drum pierced with holes and a handle used for a crank—required clothes to be hand-wrung and placed inside the drum. The drum would then be cranked to rotate over an open fire. The process smelled bad and would burn clothes when it wasn't drying them.

Before appliances for drying clothes became electric in the 1920s, the chore was attempted by passing wet clothes between hand-cranked rollers that might squeeze some of the water from the fabrics. At least it was a step up from merely beating the washed clothes against a rock in the nearby stream.

Lamp-radiator room heaters and, less frequently, electric fans, were used in some homes for drying clothes early in the century. It was an improvement over merely hanging wet clothes near a fireplace, but not much safer. By 1910, electrical power drove the washer and wringer. In 1936, an electric clothes dryer was patented.

There was another improvement by J. Ross Moore in 1938, when he invented a tumble dryer. But buyers didn't give it much of a tumble before the 1950s. Then the combination washer-dryer and the separate units became popular.

There were electrical spin dryers used in commercial laundries and in institutions such as the military, but not in many households. On the home front, the manual wringers were still used throughout World War I and II and nearly into the 1960s.

IN THE PAST, SOME SEA TRAVELERS PUT THEIR DIRTY laundry in thick cloth bags, tossed them overboard, and let the bags drag through the ocean for hours. Fun, maybe, but not so practical.

Similar to the concept of the dryer, the early clothes washers in the 1800s required mothers and daughters to place laundry in a wooden box and tumble the box via the hand-operated crank.

In 1907, the first electric washing machine, the Thor, was manufactured by Hurley Machine Corporation, Chicago. That year Maytag built its first clothes washer. It surely would have made a much appreciated gift, although not a very romantic one, for Mother's Day, which was instituted that year by a Miss Anna Jarvis of Philadelphia.

In 1915, a gasoline-powered washer was created by Maytag. Three years later, Launderette introduced a washer that used centrifugal force to extract water from clothes. Many of the early clothes washers had to be manually filled with water and manually drained. Automatic the first washers were not. Truly automatic clothes washers weren't made available until 1937, by Bendix brand. Although Bendix's washers needed to be anchored to the floor to keep from wobbling out the door during the wash cycle, that didn't keep Bendix and other brands from constantly improving the clothes washer.

In 1947, General Electric revealed the first top-loading automatic washer, different cycles were added for various fabrics, buttons replaced knobs, and by the 1950s, automatic washers sales bypassed electric washer sales for the very first time.

IN THE 1920S, EARLY FORMS OF FABRIC SOFTENERS were developed to soften the roughness of dried cotton, felt especially on diapers, underwear, and handkerchiefs. The method of dying cotton used in the early twentieth century left the products harsh to the touch, but with this new innovation, the fabric could easily be softened. The first fabric softeners were oily; people used to soak their dyed cotton clothes in a slippery mixture of water and olive oil, tallow oil, or corn oil.

But due to developments in chemistry, more effective fabric softeners were created, hitting the commercial market and becoming popular not only in the mills, but in homes. The chemicals in the early fabric softeners merely settled on the surface of fabric. Modern formulas of fabric softener allow the softening agents to actually absorb into the fabric, improving the feel and effect of the textile.

Thankfully, we've moved away from oil and toward surfactants (surface-acting agents or—to put it simply—wetting agents) that lubricate the fabric's fibers, and neutralize the fibers' charges (goodbye static cling). But the market is still changing as various companies fight to develop and maintain superior softness.

THE IRON IS ONE OF THE EVERYDAY ITEMS THAT IS NAMED after the very matter it is made of, such as dinnerware that is referred to as the family's "silver" or a jockey's shirt being called his "silks."

Clothing maintenance began as, and has remained, a distinguishing characteristic of class and care. In Greece during the fourth century BCE, slaves and servants labored over the clothing of the wealthy, perfecting it with heated metal bars. By properly folding the cloth and rolling hot bars over the folds, slaves produced pleats in the robes of their masters. This social distinguisher, and the means of creating it, spread to the Roman Empire around 200 BCE. They began using metal mallets to pound wrinkles out of, and pleats into, their fabrics. The difficult process of producing pleated cloth made the effect even more coveted by the wealthy. Both the Vikings and the ancient Chinese utilized similar methods to pleat and smooth their clothing. The Chinese placed hot sand and coal within an open compartment attached to a handle and flat bottom. In the tenth century, the Vikings used a mushroom-shaped glass-iron "smoother" heated over steam.

More contemporary irons found their way into European homes around the 1300s. First came the flatiron. True to name, it was comprised of a flat piece of iron connected to a metal handle. Heated over fire, it could be applied to clothing much like the modern day iron, smoothing the fabric until it cooled.

Although more efficient than pounding cloth with a mallet, the flatiron had some drawbacks. Flecks of ash from the fire often landed upon the flatiron, streaking clothing with black marks. Although poorer families continued to use this occasionally problematic method, by the fifteenth century, most wealthy homes possessed a hot box. Also known as a "slug iron," this contraption consisted of a handle attached to a metal box, which held hot coals or bricks. The box eliminated the problem of soot. In the nineteenth century, cast-iron stoves made flatirons cleaner and more convenient. The iron could be heated on the stove top instead of by fire.

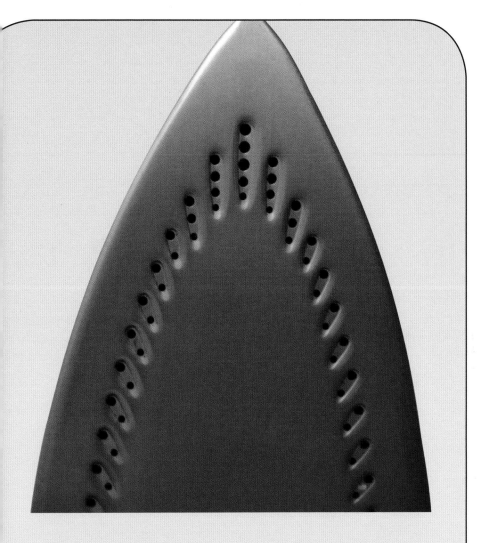

With these and other improvements, including the introduction of the wooden handle thereby reducing the danger of burning the user's hand, flatirons dominated iron technology for hundreds of years, until electricity became available in homes.

In 1882, a year that saw several practical applications of electricity, a patent for an electric iron was obtained by Henry W. Seely of New York. The device didn't take on additional significant features until 1924, when Westinghouse introduced the first automatic electric iron.

IN 1790, THOMAS SAINT, A CARPENTER IN THE PARISH of St. Sepulchre, London, developed a sewing machine with the ability to move the up-and-down action of the needle through leather (later cloth).

In 1829, a tailor named Barthelémy Thimmonier, who lived in the Rhóne village of Amplepuis, built a prototype for a sewing machine. In 1831, a military uniform maker in Paris ordered eighty of the machines. However, tailors were threatened by the new technology. Thinking their jobs were at stake, they joined together to burn down Thimmonier's garment shop, destroying all but one of the machines. Thimmonier used the surviving model to demonstrate it as entertainment for a fee.

Thimmonier made and sold inexpensive wood models of his machine and, in 1845, a French businessman successfully built a machine using all metal parts. But once again a mob destroyed the machines during a broader social revolution in 1848.

Meanwhile, back in America, the technique by which the needle moved through the cloth was eventually patented in 1851 by Isaac Singer of Boston, whose name became nearly synonymous with the sewing machine. Singer's domestic sewing machine was often simply called "a Singer," as in, "I'll sew that on my Singer."

The early machines were manually driven. Foot pedals turned the main shaft that caused the needle to pulsate through fabric as it stitched a series of loops of thread. In 1889, the Singer plant at Elizabethport, New Jersey, produced the first electrically-driven machine for sewing.

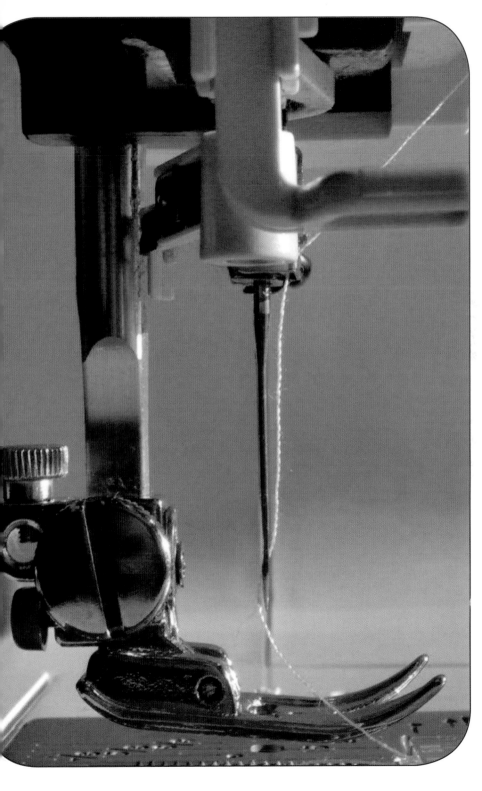

Living Room

A DEVICE THAT COULD ANSWER THE PHONE AND GIVE THE caller an outgoing message, as well as receive the caller's incoming message, was invented in 1935 by Benjamin Thornton. Primarily a machine for office use, Thorton's invention was first marketed in the 1940s.

Dozens of years passed in which Americans missed plenty of telephone calls. Then, in 1974, the importance of recording devices was dramatized by President Richard Nixon's use and misuse of them. They became more popular in the home—after all, something had to take calls while nearly eight million Americans were away: They were trying to forget Watergate and left their homes to bowl, the most popular away-from-home activity at the time. It would not be long before the sounds of a busy office could be played in certain novelty phone booths so as to assure the listener that the caller was indeed hard at work.

By the 1980s, small and friendly answering appliances became standard in the home. They remain useful and popular, even as e-mail carries the burden of most messaging today. Today's answering machine might greet its owner by saying, "While you were out, your computer called and said 'you have mail.'"

AFTER THE SUMMER VACATION, A TEACHER ASKED HER class to raise their hands if they had spent some time with Mark Twain, Charles Dickens, or Edgar Allan Poe, as they were told to do when school had ended two months before. She asked one of the students why he didn't raise his hand. The boy explained, "Our cable wasn't working."

According to the National Cable Television Association, cable TV originated in the 1940s as a way to overcome poor reception of over-the-air TV signals, especially in remote areas. Subscribers' sets were wired to antennas on high points in the areas. There were only 14,000 cable subscribers in 1952. Ten years later there were 850,000.

In the early 1970s, the Federal Communications Commission limited the cable operators to offering only movies, sporting events, and syndicated programming. However, the industry grew greatly as a result of the success achieved during the pioneering years in the area of satellite communications. In 1972, Home Box Office began. It was the country's first pay-TV network. According to the NCTA, this venture led to the creation of a national satellite distribution system that then led to cable program networks.

Between 1984 and 1992, the cable industry spent billions of dollars installing cable networks throughout the country. The NCTA states that "this was the largest private construction project since World War II."

By 1989, more than half (57 percent) of households had cable TV. Cable has undoubtedly greatly influenced the lifestyle of Americans.

DURING THE LATE EIGHTEENTH CENTURY, EARLY MOTION picture cameras were not as user-friendly as they appear to be in the movies showing them in action. The cumbersome devices required experienced operators with the skill to turn the cranks required to operate them at a consistent forty-eight frames per second. It could also be tricky to load the film into the temperamental gears and to unravel film that could jam within the gears. Not many households had them.

Cameras became more common in the 1920s once they were motorized. In 1932, 8mm film became widely available. However, it did not replace 16mm film as the amateur's choice for "home movies" before the 1950s. It was then that the Japanese company Canon began to take aim against the dominance of the giant film companies Kodak and Path.

The video camera, or camcorder, designed for home or amateur use, first appeared in Japan. During the 1980s it began to appear nearly everywhere, not long after professional versions revolutionized the video industry.

In 1975, Sony came out with its Betamax, a video cassette recorder that made it possible to record television programs. It had no competition for nearly two years until VHS format began and dominated the market. From coast to coast, consumers easily mastered the skill of recording programs off of their televisions, but setting the clock on the VCR seemed to be the biggest challenge.

(Q): How many engineers does it take to set a VCR clock?
(A): 12:00, 12:00, 12:00, 12:00

UNSURPRISINGLY, THE CHAIR HAS A LONG AND ANCIENT history. The early chairs were reserved for nobility, kings, bishops, lords, and other high-ranking individuals. (If you're wondering, everyone else sat on benches or stools). The oldest chairs existing today come from Egypt, the most ancient dating back to 2500 BCE, discovered in the tomb of Queen Hetepheres. In Egypt, chairs signified royalty. Even in hieroglyphics, the symbol for "dignitary" pictures a man seated atop this imperial piece of furniture. Because of the high status of their users, chairs were often carved from luxurious materials such as ivory and quality wood. Similarly, in ancient Rome and Greece they were carved from marble and reserved for only the wealthy and powerful. The tradition of majestic seating continued until the sixteenth century, when chairs became more common in humbler homes. However, usually only the head of the household sat in the noble seat. These chairs were rarely cushioned, but by the mid-seventeenth century, even common folk were able to rest comfortably in upholstered chairs.

During the eighteenth century, when furniture production had not yet picked up, carpenters or able craftsmen began incorporating curves into the chair. This process was significant because the curves were usually sawed out of a solid piece of wood. This piece of furniture finally became a regular in homes and castles alike, providing both the royal and average man with a seat.

THE MP3 PLAYER IS THE MOST RECENT IN A LONG LINE of music playback formats from Edison's phonograph cylinders, to vinyl discs, 8-track tapes, cassette tapes, and compact discs.

MP3 stands for MPEG Audio Layer III and MPEG is the acronym for Moving Picture Experts Group, a group that has developed compression systems for video data, DVD movies, HDTV, and digital satellite systems.

Karlheinz Brandenburg had high hopes for a PhD in Electrical Engineering. Inspired by the research he had done with Dieter Seitzer at the University of Erlangen in Nuremberg, Brandenburg began his thesis on the compression and transfer of music. In an interview with BBC News, Brandenburg recalled, "In 1988 some-body asked me what will become of this, and I said it could just end up in libraries like so many other PhD theses." Little did he know that a little more than a decade later, he would be known as the "Father of the MP3."

In 1977, Dieter Seitzer came up with the idea to transfer music over a telephone line. Although he didn't receive any grant money to aid his research, he established a team of talented individu-als to help him explore this realm of music compression, led by Brandenburg. Using creative methods and a thorough knowledge of math and engineering, they began to lay the foundation for a revolutionary experience of music.

After finishing his thesis in 1989, Brandenburg joined the pres-tigious Faunhofer Institute to further his research and develop-ment. The project experienced many technical drawbacks with their fledgling software, but through multiple tests and increased knowledge of the field, their project evolved steadily. In order to catch glitches and problems in the music system, Brandenburg used a recording of Suzanne Vega's "Tom's Diner." The simplicity and monophonic nature of the song made it easy to detect any compression issue interrupting the smooth quality of the song. The Fraunhofer Institue received a German patent for the MP3 in

1989, and ten years later, the first portable MP3 players appeared on the market and are now found in the pockets, purses, or hands of music lovers all over the world.

THE HUMAN DRIVE TO DEVELOP NEW TECHNOLOGIES CAN often be seen as an effort to extend an individual beyond his or her immediate reach—literally. The remote control is such a device that has multiplied possibilities, for both good and ill.

The remote, controller, or "clicker" as it is sometimes referred to, sends out pulses of infrared light that are codes for a specific function, like "turn on" or "turn down." The infrared receiver in the TV or stereo decodes the pulses of light into data that the device's microprocessor, or small computer, can understand. The microprocessor then carries out the corresponding command.

Remote control devices date back to the late nineteenth century with the work of the Serbian-American inventor Nicola Tesla. Tesla developed systems for wireless radio-frequency remote control of vessels and vehicles. During World War I, the German navy used this same radio-frequency remote technology to crash unmanned boats into Allied boats. In World War II, remote controls detonated bombs for the first time. Remote controls are still used to detonate bombs, with horrific consequences, but they are also used to control satellites, spaceships, and model airplanes.

The remote control has had its widest impact in the home. It has come to dominate the way people interact with their home electronics and with each other. Many would sooner search for the forever mislaid remote than get up to change the channel.

THE WORD SOFA COMES FROM AN ARABIC WORD FOR *dais*. *Dais* is the Middle English word for platform, which is from the Old French for platform. It then can be said that a sofa is a raised platform for honored guests or speakers.

The sofa has been a staple of furniture and relaxation for centuries. In the time of Romans and Ancient Egyptians, the sofa began as a cold hard bench, oftentimes wrapped in decorative fabric. In Roman society, men would stretch out on sofas while eating, the cushiony furniture arranged around a table. For most of history, the sofa was a product of wealth and status not often experienced by women or the poor.

In the eighteenth century, the Germans and English placed padding such as horsehair or dried sea moss under the fabric for extra comfort. Backrests and armrests were later introduced by the Italians, and by 1850, coil springs had been invented. Over time, contributions from different countries molded the bench into the sofa we know and love to sit on today.

Early experiments to present sound stereophonically include a device patented by a Frenchman named Clement Ader in 1881. It was used to make telephonic sounds more lifelike. It was one of several innovations related to the telephone at the time. For instance, the first telephone directory in Britain, with the names of 255 people who had phones, had appeared just the year before. Across the pond, forty-five years later, the first stereophonic radio broadcast aired from WPAJ, New Haven, Connecticut. It was "The Roaring Twenties," and the music-loving public was ready for improvements in its recordings of music. In 1926, sound on film had already been experienced by moviegoers in both the States and Britain.

In 1932, a stereophonic sound process was patented in France. Three years later, the first film to coordinate with an external stereophonic sound accompaniment was the epic *Napoleon* (a 1927 silent film). However, the music was not coordinated with the showing of the film for the entire production, which was a remarkable eight hours.

In Britain, a recorder using tape instead of wire, the ancestor of the tape recorder, was used in 1929 for synchronizing sound to film. The first stereo musical accompaniment for film was developed by the animator Walt Disney, backed by RCA. It was first used for the soundtrack of the 1941 feature-length animated film *Fantasia*. Yet, stereophonic tapes were not marketed until 1955. The stereo effect is routinely expected today in audio products. Some speaker systems have improved so much that the effect is truly realistic.

THE IDEA OF CATCHING AN IMAGE IN TIME HAS EXISTED from the first paintings on the walls of caves in France. This idea eventually grew to include the camera. A series of scientific discoveries led to the initial development of a device that could capture a moment exactly, including Sir Isaac Newton's understanding of a color spectrum within white light in 1664, but it would take longer than a flash to develop the first efficient and working cameras. In 1814, Joseph Nicéphore Niépce of France succeeded in capturing the first photographic image. Although it faded quickly, inventors could picture the potential.

The first patent issued for a camera was in 1840 to Alexander Wolcott. In his camera, he used a concave mirror to reflect the image onto photographic film. Many innovations and modifications would occur, including the invention of the still camera in 1913, the device that led the way for the digital camera and other photography developments. As automobiles became more plentiful and inexpensive early in the twentieth century because of the assembly line approach to production, cameras also decreased dramatically in price and improved in quality.

By 1930, exciting new innovations, especially with imported German cameras, made amateur candid photography immensely popular. Photoflash bulbs appeared for the first time that year. In 1947, the soft, whirling sound of the Polaroid Camera, which was invented by Edwin Land, was heard around the world. Mid-century photographers were amazed at how it developed and printed a picture in one minute. For a while few people saw anything negative about it except, of course, that it produced no negative.

Nikon's 35mm camera first appeared in 1948. The next major milestone in photography was the appearance of Kodak's Instamatic camera with film cartridge in 1963.

Aʟᴛʜᴏᴜɢʜ ᴛʜᴇ ᴄᴏᴜʀᴛs ʜᴀᴠᴇ sᴘᴏᴋᴇɴ, ᴛʜᴇ ꜰɪʀsᴛ telephone has yet to be unanimously proclaimed as such.

The name that most commonly rings a bell regarding the invention of the telephone is Alexander Graham Bell, a Scotsman dedicated to transmitting speech through electricity. Bell and his assistant, Thomas Watson, explored the possibility of a device that would enable two people to speak through electricity without the telegraph. The two worked on their two projects side-by-side, and in 1876 came success: Bell spoke into the instrument to Watson in the next room. "Mr. Watson" he said, "Come here. I want to see you." On the 14th of February 1876, Bell filed for his patent.

A less familiar story is that of Antonio Meucci. In 2002, the United States Congress voted to give Meucii, an Italian immigrant to Cuba and then New York, claim over the invention of the telephone.

He created and perfected the telephone from 1849–1871. The Italian-English barrier and his wife's paralysis in 1855 made the inventor push for a better means of communication with his wife. He rigged a system of wires throughout his home, but was unable to afford the patent. A steamboat accident left him bedridden in the hospital. Needing money to care for her husband, Meucci's wife sold the first telephone prototype to a secondhand store.

Later, Meucci created an even better model for electric communication and in 1871 he filed for a one-year renewable intent to file a patent. After 1873, however, he could no longer afford to fulfill his dream. Meucci attempted once more to receive backing, sending detailed data and his materials to Western Union. When he didn't hear back, he asked for his things to be returned. Western Union said they had lost the poor Italian's items. However, the telephone would soon be seen again by Western Union eyes, for two years later, the man sharing a

laboratory with Meucci filed for his own patent. That man was Alexander Bell.

Although Meucci attempted to sue for rights over the telephone, he died in 1889 before the case was resolved. With his death, his story disappeared. Finally, 113 years after his death, the Italian got the credit he deserved from the United States Congress. He is now recognized as the inventor of the telephone.

In 1900 America, the one million phones in use nationwide averaged out to less than one phone for fifty adults. By 1962, 80 percent of American households had a phone in their home.

THE FIRST VIDEO GAME CAME TO BE IN 1958. IT WAS called Tennis for Two, and it was invented by a man named William Higinbotham. Higinbotham's game, created and played on a Brookhaven National Laboratory oscilloscope, had the same features found in today's video games.

Tennis for Two was run by an analog computer and took Higginbotham and technical specialist, Robert V. Dvorak, only three weeks to put together.

Four years later, *Spacewar!,* an electronic game for a specific computer (the PDP-1), was developed by Steven Russell, an MIT computer programmer, who, along with other MIT hackers, were intrigued by computer developments at their time. *Spacewar!* was intended only for other programmers to play.

Within ten years, many commercial electronic games exploded into amusement arcades nationwide, addicting thousands of fans and causing them to stand for hours, mesmerized by fantastic images on large machines with challenging tasks requiring skill, experience, and many coins. Like the jukeboxes of the 1930s and 1940s, video machines soon invaded various kinds of commercial spaces, besides the arcades dedicated to them.

The first coin arcade-style game, albeit not a big favorite, was created by Nolan Bushnell in 1971 and called *Computer Space*.

In 1972, Atari served up *Pong*, an electronic tennis-like game that became for a while nearly as popular as the real game it was imitating. Within two years, a home version of *Pong* and other successful games such as *Odyssey* from *Magnavox* invaded homes. Television and electronic game-playing was about to explode in the home as it did in the arcades.

Several outstanding games surfaced among the various types that were popular in the 1990s. *SimCity* was distinctive and original, allowing the participant to create his or her own virtual city. More in the mainstream were the games created for Nintendo's Game Boy and its color version. It was matched in quality by the same company's internationally popular Pokémon game. The handheld

148

devices were smaller and could easily be taken anywhere—yes, even classrooms, concert halls, and office cubicles.

Interactivity became more innovative and technically advanced as games entered the twenty-first century with remote controls, mats for the player to dance on, and gloves to wear in order to control and interact with even more advanced screen images.

Nursery

THE ANCIENT ART OF NOURISHING A BABY WAS ALWAYS done by breastfeeding. Either the baby's mother or a wet nurse would nurse the infant. If this wasn't possible, "feeding vessels" were used—kind of an old-fashioned pre-bottle. The oldest surviving one is from 2000 BCE. These were made from wood or, later, ceramic. There was no rubber, so sometimes animal hides were used to help get the liquid into the baby's mouth.

In 1841, Charles M. Windship patented the first glass nursing bottle, and four years later, the first rubber teat was patented by Elijah Pratt of New York. However, a practical artificial teat for baby bottles was not perfected until the twentieth century. Early baby bottles were very different from what we're used to today. Some were made of clear or aqua-colored glass; many were made to lie on their sides, as opposed to upright; and some baby bottles have even been shaped like a baby's head!

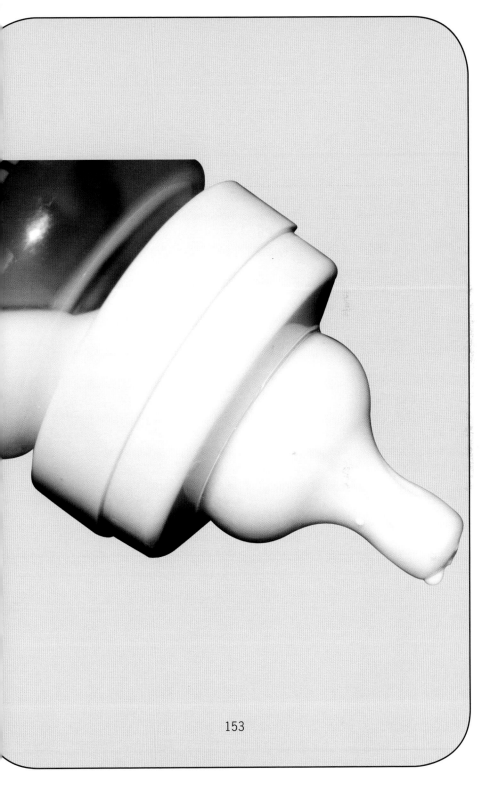

AS IT BECAME NOT ONLY POSSIBLE BUT EASY IN THE 1980S to localize simple radio communication, battery-powered, portable sending and receiving technology (e.g., walkie-talkies) flourished.

Eventually, in the late 1990s, this technology was patented, brought indoors by a woman named Cynthia L. Altenhofen, and applied to the needs of parents to watch over their newborns without having to take up permanent residence in the nursery.

With a sending unit near the baby and a receiving unit near the parents, babysitting became easier and more efficient. Some units can play baby's favorite lullabies (maybe a recording sung by the parents), and may have a nightlight, room temperature display, or clock to help baby be on time for his or her doctor's appointments. Some modern models signal the receiver with vibrations like a phone pager does. While quiet privacy is provided for baby, the parents can go elsewhere to make another baby.

Serious uses of baby monitors include their help in preventing Sudden Infant Death Syndrome (SIDS), or "crib" death. Especially helpful in that regard are newer units that include closed-circuit TV that can send live images to the receiving sets throughout the house.

Baby monitors have come a long way since the basic walkie-talkie days. Now there are video monitors, movement sensors, flat-panel color video and sound monitors, and long-distance monitors. The Fisher Price Sounds 'n Lights monitor even boasts dual receivers for adults who have difficulty hearing.

DEVELOPED IN THE LATE 1970S, BABY WIPES ARE strongly believed to have been introduced to us by the same people who brought us Kleenex and Huggies—the Kimberly-Clark Corporation.

Today, of course, there are many smaller brands and variations of wipes that address the universal little human condition. (The human is little, not the condition.)

Today's diapers and wipes are friendlier to the environment—and to baby—than earlier versions. There are even how-to directions online for making baby wipes at home. Doing so is less expensive than using commercial wipes and, depending on the paper towels and baby shampoo used, can be even less harmful to the baby. However, the choices are not so simple: Environmentally-minded individuals are switching paper towels for cloth.

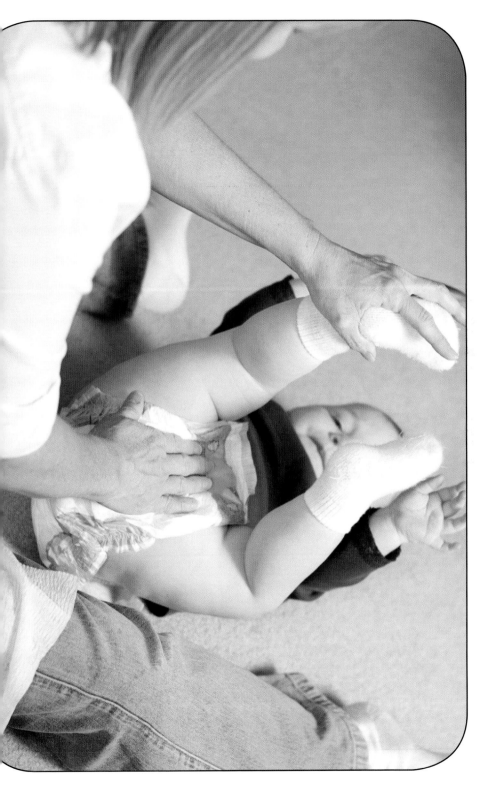

CENTURIES OF PARENTS, OLDER SIBLINGS, BABYSITTERS, and next-door neighbors understand the difficulty of lulling a cranky baby to sleep, and everyone has his or her own methods. For example, the ancient Mayans used twine hammocks to coax their infants' eyes closed. But in most cultures, babies commonly slept in the same bed as their parents until the early 1400s. In the fifteenth century, the Catholic Church encouraged mothers to sleep separately from their infants to avoid unknowingly suffocating them at night. In response, the crib was invented.

The earliest form of the crib is the rocker, or cradle. Derived from a hollowed-out half-log, they were often placed on a frame, allowing them to rock slightly. This new invention was placed next to the bed, easily accessible to the mother and father. The royal infants of the time, however, had nicer accommodations. A Gothic cradle from the end of the fifteenth century is the earliest example found of the modern cradle. When children grew too large for a half-log bed, they slept in a makeshift wooden box on the floor, which could easily slide under the main bed to conserve space during the day. By the 1800s, with more space available, these larger baby beds became taller, more permanent, and more ornate. The contemporary crib was born, quickly becoming an important artifact in every household.

IN MIDDLE ENGLISH, THE WORD *DIAPER* REFERS TO "A patterned fabric," whereas in Medieval Greek, *dia* (thoroughly) and *aspros* (white) referred to pure white. That would explain a dictionary's surprisingly specific description of a diaper as "a white cotton or linen fabric patterned with small diamond-shaped figures."

Cloth diapers made their debut in 1887. A mother, Maria Allen, invented the first cloth diaper. The first diapers were secured by buttons, needles, or straight pins (obviously the safety pin had not yet been invented) and were constantly added to the dirty laundry pile. Thankfully, much later (1946, to be precise), another mother, named Marion Donovan, invented what was called "the Boater," also known as the first disposable diaper. The first disposable diaper was made out of shower-curtain plastic that held Allen's cloth diaper. The diaper was evolving, slowly but surely.

In 1952, publications including *Modern Romances* magazine touted the diaper as the greatest single indicator of America's continued growth. The average American household had 3.37 and babies and the market was booming as unemployment was at only 3 percent. Improvements were made as competition increased.

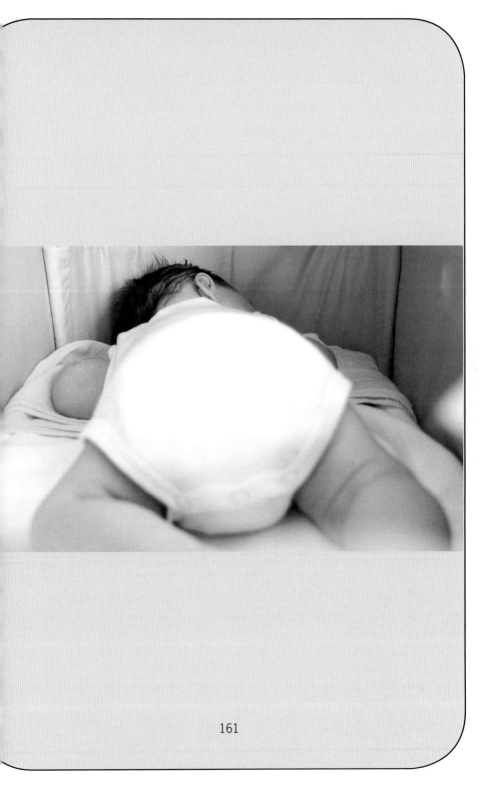

Office

Historians might enjoy pointing out that the ancient abacus can be used to achieve many of the basic calculations of the electronic calculator. And a person need not be highly skilled or even exceptionally intelligent to get the correct results with a calculator. The instrument's development over the years has expanded its capabilities and enhanced its user-friendliness.

In 1623, in Rubingen, Germany, Wilhelm Schickard invented an adding machine. Previously, the simple abacus was helpful in doing calculations, but Schickard's machine, rather than the user, did most of the work. Maybe Schickard was moved to invent his machine because the previous year, a slide rule was invented by Englishman Reverend William Oughtred of Albury, Surrey, England. As with the abacus, the user had to be actively involved in the mental calculations.

Eager to make their lives even easier and much more profitable, Americans brought a business sense dimension to previous inventions in 1888, when a keyboard was added to the adding machine concept. It was marketed by Door Felt, Chicago, and was first sold to Equitable Gas Light and Fuel Company. The first adding machine also appeared in Chicago the next year with the Comptograph, made by Felt and Tarrant, Chicago. It was first sold to Manufacturer's National Bank in Pittsburgh.

In 1960, the digital display for pocket calculators first appeared. The British inventor Clive Sinclair launched the first pocket calculator that utilized this new technology in 1972. It was called the "Executive." The most notable feature at the time was the light-emitting diode (LED) display. That same year Hewlett-Packard came out with an innovative version, as did Sharp of Japan the following year.

Today's calculators have features that mathematicians in the abacus days wouldn't have imagined—liquid crystal displays, a sleek credit-card size, and solar- and dual-powered options.

To ERR IS HUMAN; TO ERASE DIVINE? EVERYONE MAKES mistakes in life. Wouldn't it be nice to be able to simply "white them out"? Happily, mistakes on paper are more easily remedied.

There was a time when mistakes had to be corrected by hand on hard copy. There was no delete key. But erasing the error with a simple eraser left too much smudge or made a hole in the paper. So, correction fluid, also known as Wite-Out or Liquid Paper, a thin opaque fluid that is applied to written, typewritten, or printed text, was developed.

Bette Nesmith Graham, a secretary and single mother, wanted to be an artist. She would secretly take white paint and her watercolor paintbrush to work with her in the 1950s. While Graham painted over her typing mistakes, her coworkers quickly took notice and Graham gladly shared her solution. First called "Mistake Out" (later renamed "Liquid Paper") during its early marketing in 1956, the correction liquid sold successfully and quickly. Graham's business was worth one million dollars by 1967. And the rest is history.

Today, these liquid correctors consist of titanium dioxide, a polymeric film and a solvent. The titanium renders the solution opaque, and because it does not absorb a lot of visual light, it produces a predominantly white color.

Although the titanium oxide, making up 40 to 60 percent of the fluid, actually covers the error, the polymer material, making up between 5 and 15 percent of the fluid, is needed to make it stick to the paper when it dries.

Finally the solvent or thinner is added to control the viscosity and drying time of the correction fluid. Once applied, the solvent works by diluting the formula and quickly evaporating to leave a dried film.

PICTURE THE LOWLY BUT UBIQUITOUS LITTLE PAPER CLIP, a thin bit of bent wire venerated as a national symbol of unity against the occupation of Norway during World War II, yet also found littering the floors of countless offices worldwide. What would mankind do without it? What did mankind do before this gem of an invention changed everyone's life?

Mostly people used straight pins, string, and eyelets, all methods requiring that holes be made in the paper, which more often than not resulted in damage to the document. The evolution from straight pin to paper clip appears to follow Henry Petroski's notion of "form following failure." Pins fell out, damaged the paper, or injured the user, so they were replaced as soon as materials and manufacturing capacity caught up with human imagination.

The first of a myriad of variations on the theme was patented by Samuel Fay in 1867. Named the "Ticket Fastener," his product was originally meant to attach tickets to fine fabrics without creating the same damaging holes as the commonly used pins, the same theory behind the invention of the paper clip. Ten years later, Erlman J. Wright patented his "Wright Paper Clip," meant specifically to fasten loose papers and documents. Many similar clips followed, but it wasn't until the GEM paper clip of 1899 that the world witnessed the paper clip design we still use today. Patented by William D. Middlebrook of Connecticut, the GEM was named after the oval-within-oval design of the clip.

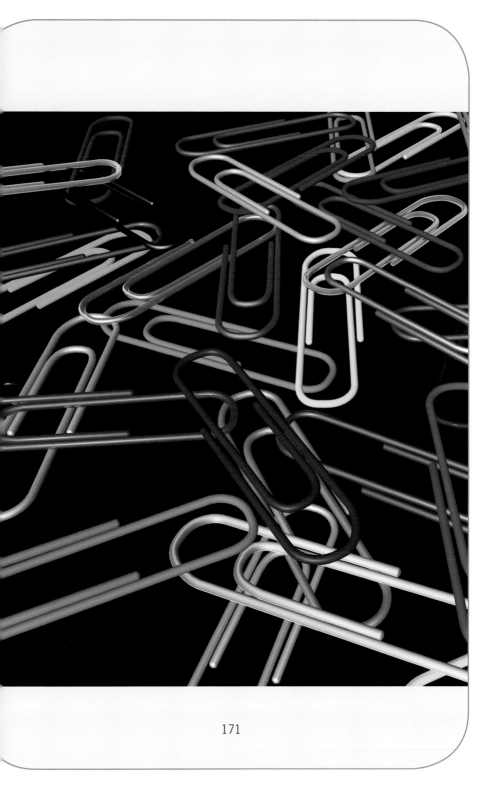

EARLY SCRIBES USED A STYLUS ON CLAY TABLETS. LATER, reeds were often cut to points and dipped in soot. The first quill pens were made in about 500 BCE and were still in use in the seventeenth and eighteenth centuries. Quill pens were made from swan or goose feathers and cut into a point at one end to make a nib. Dipped in ink, it was a slow and laborious way of writing which required the constant sharpening of the quill.

In 1656, Dutch travelers saw some of the earliest pens that contained their own ink. In 1663, the writer Samuel Pepys received a reservoir pen from William Coventry. The diarist used the innovative new tool when writing his sermons and probably also for making entries in his diary.

The term "fountain pen" was used at least as early as 1710. However, in the 1930s and '40s many school desktops had holes in the upper right corner into which small ink bottles could be inserted. Students carefully dipped the tips of their pens into the bottle to collect some of the ink on the replaceable steel pen points.

The Compound Fountain Pen, patented in 1809, was a tube with a quill nib. The tube was capped with a cork after ink was poured into it. The user could squeeze the tube as needed to deliver ink to the pen point. The device was good enough for Lewis Edison Waterman, an insurance salesman who in 1884 discovered, after much trial and error, a much better way to deliver ink to the pen's point. The resulting product—the first of millions of Waterman pens—was originally manufactured on a kitchen table in the back room of a small tobacco shop in Greenwich Village, Manhattan.

The ballpoint pen was even easier to use than a Waterman. It first appeared in retail markets in 1945. By 1950, the ballpoint in general was popular, yet still expensive. The top of the line in the postwar ballpoint craze was also the first to appear on campus— the aluminum Reynold's International ballpoint pen. It was 14.6 centimeter long, a stylish but not yet comfortable instrument to hold and use.

In 1960, the fiber-tip pen first appeared, soon to rival even the classic lead pencil in the inexpensive writing tool industry. It had the brand name Pentel and was manufactured in Tokyo by the Japan Stationery Company in Japan. Five years later in England, Penline fiber-tips pens began to be marketed by Mentmore of Stevenage.

WHERE WOULD MANKIND BE WITHOUT PENCILS? EVEN IN this digital age and supposedly paperless office of laptops and flash memory, we are ever-dependent on our writing instrument.

The first pencils could be considered to have been developed around 1560 out of sticks cut from high-quality natural graphite mined at Cumberland, England. These were wrapped in string or inserted into wooden tubes so the writer's hand would remain clean.

In 1795, French chemist Nicolas Conte developed the pencil we know today. By mixing clay and graphite and then placing the combination in a firing kiln, Conte realized he could make the lead either softer or harder, an important distinction for an artist, drafts-man, or student taking a standardized test. After being fired, these sticks were glued within a hollow slot while the surrounding space was filled with strips of wood. This design, with the addition of an eraser, conquered time and is still the standard design of pencils today.

These days, every living space is awash in pens and pencils, except of course when one is really needed. The little objects range from simple eraserless stubs to mechanical pencils of some sophistication, and from the ubiquitous ballpoint to the elegant fountain pens and the space-age Space Pen.

Contemporary mechanical pencils, the first of which were pat-ented by Sampson Mordan in 1822, use thin leads, made possible through advances in lead composition. Ordinary pencil lead is a mixture of graphite and clay. But at very thin (.5 millimeter) sizes they are too brittle and tend to snap too easily. Modern leads are based on carbon made from high polymer organic materials and can take a little more force and flexing without snapping.

SO WHAT'S IN A PENCIL SHARPENER? KNIVES, OF COURSE. The knife is the original pencil sharpener and probably the one still turned to when nothing else is handy.

Today we commonly reserve sharpening for our pencils; however, the original need for a tool to combat a dull point came from the first pens. From 500 BCE through the eighteenth century, quills were repeatedly dipped in ink and dragged across paper and, just like our modern-day pencils, the tips became blunt and difficult to use. Enter the penknife. With the introduction of the wooded pencil in 1795, the penknife acquired a new edge to taper. However, the many disadvantages of using a knife, including the lack of a consistently good point and the tendency to break the tip while sharpening, led to the invention of a more modern pencil sharpener.

In 1828, the French mathematician Barnard Lassimone patented the first sharpener solely for the pencil. A second Frenchman, Therry des Estwaux, came up with the manual pencil sharpener in 1847, paving the way for the sharpeners we use today. The first machines to hit the market used harsh materials such as steel or sandpaper to grind the pencil lead to a point. After 1896, the popular method was to use a series of sharp knife-like edges, known as milling cutters, to whittle down the wood. By 1915, the common grinder sharpener with spiral cutting edges (small knives) on twin cylinders contained by a cover to catch the shavings had come to dominate the sharpener market. As pencil sharpeners became more common, they became smaller and more effective. In the early 1940s, the first electric sharpener was patened by the Hammacher Schlemmer Company. Today, both manual and electric sharpeners are commonly used.

Nevertheless, hundreds of models of small handheld sharpeners continue to be produced and are often the object of collectors. These are essentially small knives set in a wide variety of housings with a conical hole in which to insert the pencil—everything from cannons, airplanes, and signs of the zodiac to a small doll head that spits the shavings out of its mouth. Ah, the human imagination!

PERSONAL COMPUTER

THE MOST IMPORTANT INVENTION IN THE SECOND HALF OF the twentieth century was the personal computer. The developments that have been and are being made right at this moment on the computer are simply remarkable.

It's important to know that the earliest computers did not sit on one's lap. They barely even sat on one's desk. The first computers were programmed with plug boards and switches. Some computers, like the John Mauchly and J. Presper Eckert's ENIAC from 1946, took up one thousand square feet of floor space. "Personal," indeed. And never mind bytes to measure speed. The ENIAC machine went as fast as five thousand operations per second.

Early computers were used by large companies, manufacturing plants, and scientific laboratories that actually had the means and reason to use such sophisticated equipment, but with few hacker exceptions, of course. (See: Video Games on page 150.) Therefore, most computer innovations went unknown to much of the public.

Just two years after the reveal of ENIAC, IBM brought us the Selective Sequence Electronic Calculator that stretched a mere twenty-five by forty feet. Despite its size, the SSEC was sleek enough to be a computer that assisted in plotting the course of the 1969 Apollo 11 flight to the moon.

In 1968, Hewlett-Packard joined the general purpose computer ranks with the HP-2115. Hewlett-Packard's model began the drift from "computer" to "personal computer" as computers became more compact, more affordable, and user-friendly.

The Apple II, released in 1977, actually resembled something we're used to today: keyboard, manual, A/C powercord, graphics (when hooked up to a television), and even a computer game!

The greatest-selling single computer model of all time was the Commodore 64. Although its sales were discontinued in 1993, it has sold more than 22 million units. When it was first released, the C64 sold for $595 and offered 64 KB of RAM and advanced graphics.

Apple's launch of the Macintosh in 1984 was the first successful mouse-driven computer with a user interface. It had the

applications most of us may remember: MacPaint and MacWrite. The Macintosh, like many of today's modern luxuries, was featured in a $1.5 million Apple commercial that debuted during—you guessed it—the Super Bowl.

Home computers would grow to the point that, by 1997, 40 percent of U.S. households had a personal computer.

Smaller components and the liquid crystal display (LCD) made computers portable. Laptops became a nearly indispensable business tool and for many people replaced the personal computer. As access to the World Wide Web became easier and friendlier, interest intensified worldwide and by 1998 there were thirty million users in the United States alone.

THE NEED FOR PERSONAL OR HOUSEHOLD PRINTERS accompanied the introduction of personal computers in the mid-1970s. The highest quality printers were close cousins to the typewriter, using similar printing technology. Images on a daisy wheel struck paper through a carbon tape, leaving the image the same as typewriters did using a ribbon. At first they were slower than an average typist's speed and certainly noisier than typewriters.

A dry printing process (aka electrophotography, aka Xerox) was invented in 1938 by Chester Carlson. The foundation of his invention would later become the blueprint for laser printers.

During the development of alternatives to the daisy wheel printers in the late 1950s, dot matrix printers were introduced for mainframe computers. Paper in accordion folds, with holes along the side edges, was fed through the printers. The process was mercifully quieter than previous printing methods.

Printing by fax machines was less noisy and faster, although such machines were needed less and less as e-mail became the more popular and economical method of delivery for many, if not most, documents.

Starting in 1981, ink jet printers appeared, with Japanese companies including Canon and its "bubble jet" ink printer leading the way. However, in 1984, the American company Hewlett-Packard became the industry leader with its ink jet printers and convenient ink cartridges.

As WHEN EMPLOYING THE WHEEL OR OTHER SIMPLE TOOLS, using the humble and simple rubber band touches a simple descendant of the rich history of technological achievement.

Around 1828, a British rubber manufacturer named Thomas Hancock provided a factory and workmen to a business that made water-resistant cloth. The new venture was in the Paris suburb of Saint-Denis. It was there in 1830 that the first woven elastic material was made. The following year a fashion magazine praised the "recent discovery" and its use in corsets, as it "substituted India rubber for elastic wires."

In 1839, the Charles Goodyear Company of Philadelphia, Pennsylvania, developed vulcanized rubber. The first commercial application of Goodyear's vulcanized rubber process was used the next year in Springfield, Massachusetts, for decorative cloth. Thomas Hancock, associated with several pioneering ventures in Great Britain, obtained a patent for a vulcanized rubber process in 1843.

In 1845, a patent was granted to Stephen Perry, a rubber manufacturer in London, for the first elastic bands described as being "for paper, letters, and so forth." Of course ingenious minds soon found many other uses for the rubber band, such as to twist and thereby drive small model airplanes for a few feet, thrilling children with flights of imagination. The bands were also both ammunition to be fired from the pointed fingers of school children, as well as to power more harmful slingshots. Schoolboys, probably less willingly, also benefited from elastic bands after 1887, when "knicker elastic" was introduced. Its purpose was to prevent insects and dirt from getting under the shortened pants by keeping the opening tightly closed as young legs ran about. It later became a fashion statement.

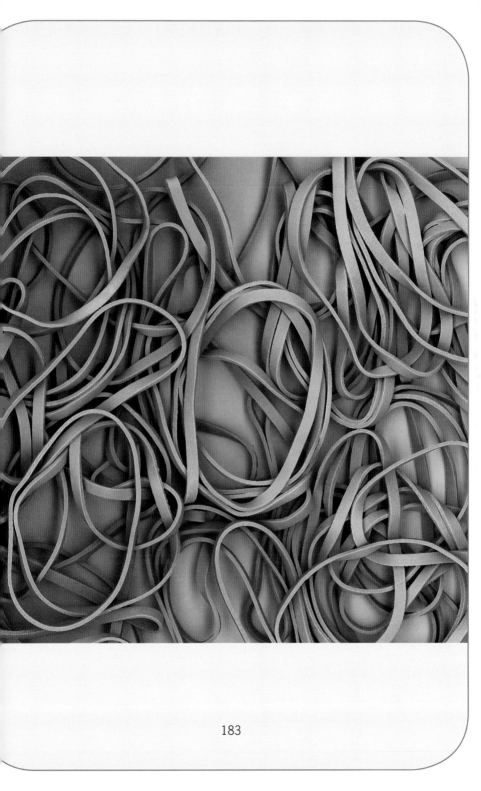

THE MAJESTY OF THE STAPLER IS OFTEN OVERLOOKED, BUT this fascinating fastening tool has been, from the beginning, fit for a king. During the reign of King Louis XV of France, his royal engineers developed the first stapler. Each individually handcrafted staple was meticulously inscribed with the insignia of the royal court. But when Louis died in 1774, the invention disappeared from the palace.

According to the Early Office Museum, in 1854, Hymen L. Lipman received a patent for his eyelet machine. It was later advertised as "The Best Paper Fastener," but his eyelet machine was not originally designed for paper. (See the article on the pencil, page 176, for more about Lipman.) Many believe that the glory of inventing the first modern stapler belongs to George W. McGill. McGill received a number of patents for various staplers and staple modifications. Credited with the bendable brass U-shape design we use today, he came up with the modern staple. A patent was obtained in England by C. H. Gould of Birmingham in 1868 for a stapler and McGill's Single-Stroke Staple Press, patented in 1879, was one of the first staplers to be found on office desks.

Although the shape stayed a staple of the U-shaped invention, the material changed from brass to wire, and from a spool to a straight pin placed in a machine and then bent.

The word "stapler" as meant here, first appears in 1887 in a filing for a U.S. patent. Before the 1920s, the terms "paper fastener" and "stapling machine" or "staple binder" were used.

Adhesives have really evolved since 4000 BC when people repaired clay pots with tree sap. Animal-derived adhesives and natural substances such as fish glue, blood, stag horn, egg whites, hide, and vegetable matter are the earliest forms of adhesives. It's safe to say that before there was tape, there was glue. Without one, we wouldn't have the other. In fact, their histories have intertwined since the year 1200 AD, when a Benedictine monk applied fish glue (which was commonly made from the air bladder of a sturgeon) to the skin of an eel and technically created the first prototype for adhesive tape.

Interestingly enough, the first instance of tape made its appearance in 1845 when a surgeon named Horace Day invented cloth-backed surgical tape. Nearly 80 years later, a Johnson and Johnson cotton buyer, Earle Dickson, followed Day's lead and invented the Band-Aid. Soon after the success of the Band-Aid, an engineer for the Minnesota Mining and Manufacturing Company, Richard Drew, invented the first non-cloth-backed tape in 1925: masking tape. Appropriately, today we know the Minnesota Mining and Manufacturing Company as 3M. Johnson and Johnson went on to alleviate a need for a waterproof tape that could be used for emergency equipment repairs during World War II: duct tape.

Outdoors

SINCE THE BEGINNING OF FIRE AND FOOD, PEOPLE HAVE been cooking over an open flame, but to "barbeque" is not simply tossing food on a fire. The process of barbequing requires meat to be slowly cooked by heat, smoke, wood, or coals. The first BBQs were documented by Spanish explorers in the Caribbean. When Columbus sailed the ocean blue late in the fifteenth century, he observed the Taino people of the Bahamas. By digging a hole in the ground and placing wooden boards above a flame, the Taino cooked their meat with the smoke from the fire below. The word "barbeque" is thought to be derived from their word "barbicoa," meaning "sacred fire pit."

In the 1800s, barbeque hit the cooking scene of the southern United States where it has since developed into a proudly celebrated tradition. As beef became a profitable business in the mid-nineteenth century, cowboys drove their cattle from the ranches to the railroads. On these journeys cowboys were provided with less than prime cuts of meat, usually brisket, which took five to seven hours to cook properly. Chewing on tough meat and sitting around the campfire led to the birth of the barbeque in the United States.

In 1897, Ellsworth B. A. Zwoyer patented the first charcoal briquette. With his son, Zwoyer built manufacturing plants in New York and Massachusettes, but these shut down during the Great Depression, taking Zwoyer's fame with them. But by 1920, Henry Ford had already usurped Zwoyer's throne, becoming King of the Briquette. Contrived with Thomas Edison, Ford created the charcoal cooking contraptions, which he later sold to E. G. Kingsford. Kingsford commercialized the product and the barbeque was mass produced for the American people.

ARE YOU IN THE DOG HOUSE? WELL, YOU COULD BE EITHER temporarily in hot water with your spouse or you could be in a small shed commonly built in the shape of a little house, intended for a dog. A doghouse, also known in England as a kennel, is a structure in which a dog is kept or can run to for shelter.

Egyptian nobility reserved mud-brick kennels for their pooches. These kennels are among the oldest doghouses ever recorded. In Chinese, Greek, and Roman societies, a dog was often considered a status symbol, so these dogs shared a roof with their owner. Lucky dogs.

When dog breeding picked up in the 1800s, so did dog shows. For the dogs, this meant travel and board within the confines of wooden crates for a large portion of their lives. During World War II, mine sniffers, messengers, and trackers called the vented wooden boxes in which they were transported their homes while on the battlefield.

Although wood is now the preferred material for doghouses (think Snoopy), plastic doghouses were introduced in the 1960s and have gained popularity.

FERTILIZERS HAVE BEEN USED TO PROMOTE PLANT AND crop growth since the advent of agriculture in ancient times. Most forms contain one or more of the three elements important to a fertile soil. Manure and guano contain nitrogen. Bones, usually ground fish bones, contain phosphorus. And wood ash contains appreciable quantities of potassium.

Although manure, ash, and bonemeal have been used to improve crops for centuries, the use of commercially produced fertilizers is one of the great innovations of the eighteenth and nineteenth centuries and it led to a revolution in agriculture in England. An increase in agricultural productivity supported a population growth which supplied the much needed labor to propel the Industrial Revolution.

Although potassium was readily available from potassium chloride mines in Germany, phosphorous and nitrogen were more difficult to obtain. In 1841, John Bennett Lawes, an English chemist, patented the process of using sulphuric acid on phosphate-containing rocks to create superphosphates which could be readily applied to the soil. By 1870, England had about eighty factories making super phosphate.

Nitrogen production was more problematic. It wasn't until the early 1900s that a German chemist, Fritz Haber, developed the technique to produce ammonia synthetically. In 1913, Carl Bosch, a German chemist working for the stock corporation, Badische Anilin- & Soda-Fabrik (BASF), developed the industrial ammonia production process used today.

These synthetic fertilizers have dramatically increased crop yields since the early twentieth century and led to another agricultural revolution commonly called the Green Revolution. Since World War II agricultural production and yield fueled by modern fertilizers has supported a significant increase in global population.

THE ORIGINAL JUNGLE GYM WAS INVENTED IN 1920 BY Sebastian Hintonm, a lawyer from Chicago. His wife Carmelita Chase, a progressive educator, ran a nursery school there. Hinton's patent consisted of a large metal, cubic-shaped pyramid of metal pipes. He later sold it across the United States under the trademark name "Jungle Gym."

In his patent application Hinton describes climbing as a complete form of exercise and designed the Jungle Gym with that in mind. He saw it as "the natural method of locomotion which the evolutionary ancestors of the human race were designed to practice, and therefore ideally suited for children."

It was Hinton's father, a mathematician, who first constructed a similar apparatus out of bamboo in Japan. He wanted his children to achieve an understanding of 3-D space in a game called "three dimensional tag." Mimicking a Cartesian-coordinate system in mathematics in which numbers for the x, y, and z axes were called out, each child tried to be the first to grasp the indicated junction in this pyramid of cubes.

Many of the original metal jungle gyms have all but disappeared from playgrounds, removed as a result of head injuries, as well as bruises, sprains, and fractures suffered by children through falls and improper swinging on the bars. It is now more common to find rope constructions or rows of overhead bars, often referred to as monkey bars. These are high enough for a child to hang from, but not so high as to cause serious injury in a fall. A child can "walk," hand-over-hand, from one end to the other.

GEORGE BALLAS, AN ACCOMPLISHED DANCE INSTRUCTOR in Houston, Texas, was sick of the painstaking process of removing the weeds from around his trees and house. On top of it all, his car was dirty.

While still grim over his growing grass, Ballas went to the car wash, but he didn't pay attention to his newly polished vehicle. Instead, he focused on the giant blue spinning machines that whacked the dirt from the car's exterior. While watching the rapidly twirling fabric clean his car without damaging the paintjob, the WeedEater materialized in his mind.

In 1972, Ballas cut holes in a popcorn can, tied nylon fishing line to the outside, and attached it to a rotary brush cutter. The nylon effortlessly and efficiently cut through the weeds, and the slow tedious work of weeding transformed into a quick and easy outdoor chore. He immediately patented his product, eventually selling it to Emerson Electric. The WeedEater became a gardening necessity, dancing its way across lawns everywhere.

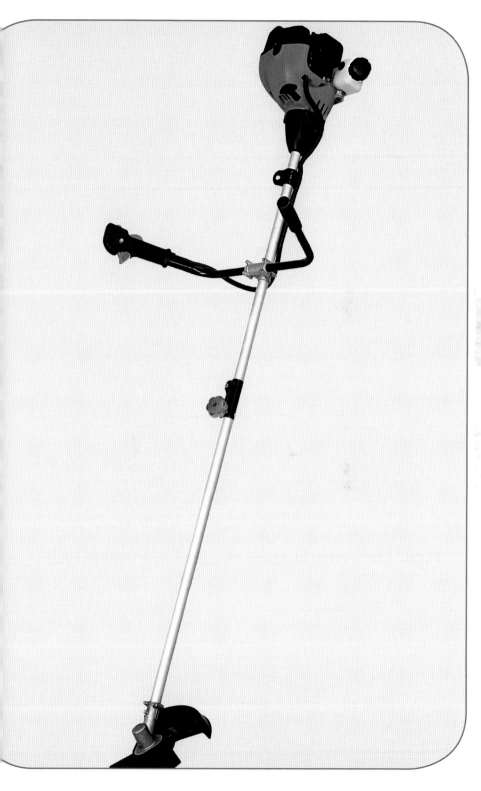

LAWNS OF PERFECTLY CUT GRASS WERE ONCE NEARLY impossible to create and maintain. But the difficulty associated with elegance added to its grandeur, especially for the French aristocracy of the eighteenth century.

The French first popularized the short grass of a manicured lawn in the 1700s, the trend quickly traveling to England where it would be absorbed into the culture. The well-kept lawn instantly became a symbol of status. Among the rich, the art of landscaping exploded. Short grass became a necessity for wealthy Europeans.

In order to achieve a manicured effect, the English and the French used the first lawn mowers: cows and sheep. Grazing animals kept the grass neatly trimmed and nicely fertilized. However, as walking through personal gardens became both a popular activity and important opportunity to show off one's wealth, the aristocrats desired an alternative method without the smelly side effects of animal lawn mowers.

Replacing the cows and sheep, servants began cutting the grass with the scythe, an instrument used for harvesting crops in Germany as early at the thirteenth century. The scythe has a long wooden handle and a curved blade that constantly needed to be sharpened as servants moved across the field. As with the first irons and pleated clothing, the ability to keep a manicured lawn clearly indicated large amounts of money. But with Edwin Budding and the invention of the mechanical lawn mower, the clean aesthetic was much easier to come by. In 1830, Budding first got the idea for this grass-getting gadget while walking around a cloth mill. He noticed a rotating blade trimming the excess fabric off newly woven cloth, giving it a clean and impeccable finish. He realized that if mounted on top of wheels, a similar machine could be used for cutting grass.

The lawn mower became popular all over Europe and America. The item was cheap and easy to use, and fresh cut grass became a staple for homes across the world.

Swinging has soared into our hearts for more than twenty-five centuries, carrying us into the past and shooting us toward the future. The first documentation of swings occurs in China as early as 475 BCE, where they were used to strengthen and train the warriors of the Shanrong tribe. It served as a military device through 221 BCE. When the Shanrong were defeated, these curious contraptions were shipped off to the central plains of China where they were first enjoyed by the royal children. Soon, swings became a staple of springtime, enjoyed by royal children, women, and the men of China.

As swinging became more popular, it also became more luxurious. Beautifully colored silk ribbons were attached to the swings, fluttering with the movement of the palace women. It also became a designated event at the Qingming Festival during the Ming dynasty from 1368 through 1644. But by this time, swinging was not only a recreation of the Chinese.

In the eighteenth century, European paintings, specifically French, depicted numerous scenes complete with rope swings. Often, the swings were loops of rope attached to tree branches, with two more ropes on either side of where the swinger sat. Most often, a woman claimed this position and the man courting her would pull on the ropes, bringing her toward him and then releasing her back.

Even before these pictures from France, statues and paintings from ancient Crete and the Far East show the joy of swinging, which has remained a favorite activity through the ages. And today, all one must do is walk to the nearest playground.

Acknowledgments

For their generous cooperation, thanks to William Kuhns, Herb Verbesey, Joe Wessling, Jerry Hein, and others with whom there is a common bond. Special thanks to Stephany Evans of Fine Print Agency and to Kathleen Go and Ann Treistman of Skyhorse Publishing.

Resources

About.Com. 2008.
<http://inventors.about.com>

"About Zenith: Remote Control." *Zenith Electronics Corporation.* 2007.
<http://www.zenith.com>

Ament, Phil. *The Great Idea Finder–Celebrating the Spirit of Innovation.*
<http://www.ideafinder.com>

Answers.Com–Online Dictionary, Encyclopedia and Much More.
<http://www.answers.com>

"Baby Wipes." *Mahalo.Com: Human-Powered Search.*
<http://www.mahalo.com>

"Big Hair Moments." *PBS.*
<http://www.pbs.org>

Blackburn, Graham. "A Short History of Beds, Cradles, and Cribs." *The Taunton Press.*
<http://www.taunton.com>

The Black Inventor Online Museum.
<http://www.blackinventor.com/>

"Bleaching," *Microsoft Encarta Online Encyclopedia.* Microsoft Corporation, 2008.
<http://encarta.msn.com>

Boese, Alex. "History of the Bathtub." *The Museum of Hoaxes.*
<http://www.museumofhoaxes.com>

Bogucki, Ed. *History of Nursing Bottles.*
<http://acif.org/past.html>

Brookhaven National Laboratory. "Video Games—Did They Begin at Brookhaven?" *Doe R & D Accomplishments.* 1981.
<http://www.osti.gov>

RESOURCES

Byers, Anthony. *Of Service: a History of Electricity in the Home*. London: Electricity Council, 1981.

Carmack, Sharon D. "Cooling Trends." *Family Tree Magazine*. <http://www.familytreemagazine.com>

Cate, Matthew. The Daily Beacon, University of Tennessee. 19 Jan. 2001. <http://dailybeacon.utk.edu>

Chapel, George L. "Gorrie's Fridge." Historical Society, Inc. <http://www.phys.ufl.edu>

"Chlorine Chemistry Division." American Chemistry Council, Inc. <http://www.americanchemistry.com>

Computer History Museum. <http://www.computerhistory.org>

Conn, Charis, and Lewis H. Laham. *The Harper's Index Book*. Vol. 3. New York: Franklin Square P, 2000.

De La Rocha, Kelly. "Parenting Innovations." *Albemarle Family*. <http://www.albemarlefamily.com>

"Dog House History." *All About Dog Houses*. <http://all-about-dog-houses.com>

Dorfman, Marjorie. "The Bath Tub." *Home is Where the Dirt is*. <http://www.housenotsobeautiful.com>

Early Office Museum. <http://www.officemuseum.com>

Edelman, Jonathan. "A Brief History of Tape." *AmbidextrousMag.org* <http://www.ambidextrousmag.org/issues/05/pdf/i5p45_46.pdf>

"Fabric Softeners." *About Cleaning Products*. <http://www.aboutcleaning.com>

"Fireplace History." *FirePlaces*. 2004 <http://www.fireplaces-fireplaces.com>

"Food Graters." *Earth Science Australia*. <http://earthsci.org>

French, Christy T. "The History of Makeup." *Authors Den*. <http://www.authorsden.com>

Gordon, Lois, and Alan Gordon. *The Columbia Chronicles of American Life*. New York: Columbia UP, 1995.

"Heating, World of Invention." *Bookrags.Com: Book Summaries, Study Guides*. <http://www.bookrags.com>

"The History of Household Wonders." *History.com*. <http://www.history.com/exhibits/modern/vacuum.html>

"History of Mattresses." *Bed Mattresses.*
   <http://www.bedmattresses.info>

"History of the Stapler." *The Antique Stapler Collector's Website.* 2004.
West Groton Company.
   <http://www.westgroton.com/staplers/index.html>

Hunt, John A. "A Short History of Soap." *The Pharmaceutical Journal*
263.7076 (1999): 985–989.
   <http://www.pharmj.com>

Hushamok™ 2007.
   <http://www.hushamok.com/history.cfm>

"Inventions: an International Affair." *InterSol, Inc.* 1996.
   <http://intersolinc.com>

Krumholz, Phillip. "Shaving and Barberiana: History of Shaving." *Heart
Technologies, Inc.* 27 Jan. 2007.
   <http://www.heart.net>

Kurtus, Ron. "Biography of King Gillette." *School for Champions.* 2008.
   <http://www.school-for-champions.com>

Lahanas, Michael. "Furniture and the Greek House." *Hellenica.*
   <http://www.mlahanas.de>

Liegey, Paul R. "Hedonic Model for DVD Players." *Bureau of Labor
Statistics.* U.S. Department of Labor.
   <http://www.bls.gov>

Merriam-Webster Online–Dictionary and Thesaurus.
   <http://www.merriam-webster.com>

"Modern Marvels: Garage Gadgets–Weed Wacker." *The History
Channel.* 2008. A&E Television Networks.
   <http://www.history.com>

"MP3 Creator Speaks Out." *BBC News.* 13 July 2003.
   <http://news.bbc.co.uk>

Mr.Coffee.com. 2008.
   <http://www.mrcoffee.com>

National Academy of Engineering. "Household Appliances Timeline."
   <http://www.greatachievements.org>

"The Not So Friendly Bottle." *The History of the Feeding Bottle.*
   <http://www.babybottle-museum.co.uk/murder.htm>

Ochs, Carol. "History of Soapmaking." *Simply Soap.* 1994.
   <http://www.simplysoap.com>

Original Hanau Suncare.
   <http://www.original-hanau-suncare.com>

RESOURCES

Panati, Charles. *Extraordinary Origins of Everyday Things.* New York: Harper & Row, 1987.

Patent Analytics and Patent Searching.
<http://www.freepatentsonline.com>

Petroski, Henry. *The Evolution of Useful Things.* New York: Vintage, 1992.

Reese, Randi. "The History of Baby Cribs." *Isnare.* 2007.
<http://www.isnare.com>

Robinson, Patrick. *The Book of Firsts.* Bramhall House, 1974.

"Roman Hypocaust." *Historic Herefordshire Online.*
<http://www.smr.herefordshire.gov.uk>

Schunck, Rebecca. *"Wallpaper History."* Wallpaper Installer.
<http://www.wallpaperinstaller.com>

"The Shaving Historical Timeline." *Quik Shave, Inc.*
<http://www.quikshave.com>

Solter, Aletha. "Should I Let My Baby Sleep with Me?" *Aware Parenting Institute.* 2001.
<http://www.awareparenting.com>

Springer, Ilene. "How the Ancient Egyptians Put Their Feet Up: Furnishings in Ancient Egypt." *Tour Egypt.* 16 Jan. 2006.
<http://www.touregypt.net>

"Toothbrush History." *Parenting Toddlers.*
<http://www.parentingtoddlers.com>

Tout, Nigel. "Clive Sinclair and the Pocket Calculator." *The International Calculator Collector.* 2003.
<http://www.vintagecalculators.com>

"Tv Trays–Tables." *Home Furnish.Com.*
<http://www.homefurnish.com>

Van Vleck, Richard. "American Grain Cradles." 1998.
<http://www.americanartifacts.com>

Wallechinsky, David, and Irving Wallace. *The People's Almanac.* Garden City, NY: Doubleday, 1975.

Webb, Pauline, and Mark Suggitt. *Gadgets and Necessities: an Encyclopedia of Household Appliances.* Santa Barbara: ABC-CLIO, 2000.

"Who Invented the Television?" *The Tech FAQ.*
<http://www.tech-faq.com>

Wikipedia.
<http://www.wikipedia.org>